READ

Quantum Leap
AD...

KNIGHTS OF...
After his latest Leap, Sam... facing a man in full armor!

PULITZER
Was Al a traitor to his country? Only Sam can
find out for sure . . .

ODYSSEY
Sam and Al must save a twelve-year-old from an uncertain
future—and himself . . .

INDEPENDENCE
In 1776, Sam's ancestor could have been a patriot or a Tory—
now he has to find out first-hand!

ANGELS UNAWARE
When Sam Leaps into a priest, Al joins him as an angel in a
quest to ease a woman's pain.

OBSESSIONS
A woman claiming to be Sam's wife threatens to turn Project
Quantum Leap into a tabloid headline.

LOCH NESS LEAP
Sam makes a monster Leap into a physicist at Loch Ness—but
the reason behind this mission is the real mystery . . .

HEAT WAVE
Sam must move fast to prove a man's innocence—while
keeping peace in a town about to be torn apart
by racial violence.

Quantum Leap
**OUT OF TIME. OUT OF BODY.
OUT OF CONTROL.**

QUANTUM LEAP
FOREKNOWLEDGE

A NOVEL BY
CHRISTOPHER DEFILIPPIS
BASED ON THE UNIVERSAL TELEVISION
SERIES *QUANTUM LEAP*
CREATED BY DONALD P. BELLISARIO

Berkley Boulevard Books, New York

If you purchased this book without a cover, you should be aware that this book is stolen property. It was reported as "unsold and destroyed" to the publisher, and neither the author nor the publisher has received any payment for this "stripped book."

Quantum Leap: Foreknowledge, a novel by Christopher DeFilippis, based on the Universal television series QUANTUM LEAP, created by Donald P. Bellisario.

QUANTUM LEAP: FOREKNOWLEDGE

A Berkley Boulevard Book / published by arrangement with
Universal Studios Publishing Rights,
a Division of Universal Studios Licensing, Inc.

PRINTING HISTORY
Berkley Boulevard edition/March 1998

All rights reserved.
Copyright © 1998 by Universal Studios Publishing Rights,
a Division of Universal Studios Licensing, Inc.
Cover art by Stephen Gardner.
This book may not be reproduced in whole
or in part, by mimeograph or any other means,
without permission. For information address:
The Berkley Publishing Group, a member of Penguin Putnam Inc.,
200 Madison Avenue, New York, New York 10016.

The Putnam Berkley World Wide Web site address is
http://www.berkley.com

Make sure to check out *PB Plug*, the science fiction/fantasy newsletter, at
http://www.pbplug.com

ISBN: 0-425-16487-X

BERKLEY BOULEVARD
Berkley Boulevard Books are published by The Berkley Publishing Group,
a member of Penguin Putnam Inc.,
200 Madison Avenue, New York, New York 10016.
BERKLEY BOULEVARD and its logo are trademarks belonging to
Berkley Publishing Corporation.

PRINTED IN THE UNITED STATES OF AMERICA

10 9 8 7 6 5 4 3 2 1

For Laura, my wife.
Your love made it possible.

ACKNOWLEDGMENTS

I want to thank the following people, without whose input this book wouldn't have been possible:

Mom & Pop, first and foremost, for giving me life then teaching me what to do with it. I think you'll see a lot of yourselves in these pages.

Michael Moeder, whose screenwriting class forced me to write the scene that evolved into this book.

Nancy Henderson, who always had the facts on hand to take care of my writing emergencies; and Nancy, who along with Ruth Calkins got me in under the wire at the first Eastleap.

Jo-Ann Wilson, for her careful reading and for providing the Geechee connection.

Tom Dunn, for editing after midnight.

Elizabeth Storm, for an invaluable source of support and e-mail procrastination.

Owen McGovern, for letting me moonlight, and the rest of the CPA staff, who had to put up with me while I was writing this.

Ginjer Buchanan, who gave me a shot and waded through three years of revisions.

And, last, Uncle Carl, who started the ball rolling when he asked me, "But what happens when the people go back?" Hope this answers the question....

Author's Note:

The events in this novel take place sometime after those chronicled in the television episode "Trilogy" (parts 1–3).

PROLOGUE

"He's coming now, Sam. It's showtime."

Sam nodded, and finished applying his lipstick. He pursed his lips, trying to look seductive.

"Forget it," Al quipped. "You look like a striped bass."

Sam snapped the compact shut and returned it to his purse. Al was right. He would have to let his eyes do the talking. *And* the low-cut dress.

There was a soft knock on the door and Colton walked in, a portrait of arrogance and power.

"Well, if it isn't the count of no account," sneered the Observer. "I said it before and I'll say it again, Sam. This is probably the jerk who invented the power tie."

Sam forced a smile. "Hello, Jonathan."

"Good evening, Ann-Marie," Colton said in a smug British accent that oozed like rancid butter.

"*'Good evening,'*" Al disgustedly mimicked. "Would you listen to this phony? I've seen cartoon characters with more soul."

Sam knew his friend was right. Ziggy had traced Colton's

lineage to Palmyra, Missouri. He was just plain Johnny Reinhold there. But the accent was part of the persona that the farm boy-turned-royalty had built for himself while insinuating his way into New York's jet set. The socialites had devoured his claim of being a minor lordling and asked for seconds. Making the scene led to contacts, and Colton's import/export business had thrived.

"I must admit I was somewhat confused when you asked to meet me in your office at so late an hour," Colton continued. His eyes wandered up and down Sam's body until they hovered at breast level. "But it's becoming increasingly clear."

Slimeball! Sam thought. But he willed the smile to remain. This was the reaction he was looking for, after all. It was the only way left that he could think of to pull himself out of this disastrous Leap.

"Is it?" Sam asked, approaching Colton slowly. Colton smiled and sat down on the arm of one of the matched set of heavy wooden chairs that faced an intricately carved antique Chinese desk. "Whatever do you mean?" Sam played with the idea of fluttering his eyelashes but dismissed it. *I can't lay it on too thick....*

Colton reached out, drawing Sam closer and rubbing the back of Sam's neck. "Yick!" said Al.

"Now"—Colton shot his hand up and grabbed a fistful of Sam's hair—"maybe you can tell me what this is *really* all about."

Sam fought to pull away. "Give him a screaming bonsai chop in the throat, Sam!" yelled the Observer. Sam was inclined to agree, but at the first hint of movement, Colton pulled harder. A sudden twist would leave Sam with a broken neck.

"Taking self-defense classes?" asked Colton with contempt. "How very *liberated*. But let's not forget who calls the shots around here, shall we? Now, what do you want?"

"Diamonds," said Sam through clenched teeth. "I have some clients very interested in rare gems. Price is no object."

The last seemed to get through to Colton. He eased his grip a little. Sam knocked Colton's arm away and broke free.

"Gems, eh?" asked Colton. "And who are these clients?"

"That's not important. What matters is that they're ready to do business. And for that, I need you."

Colton feigned surprise. "Our self-reliant little spitfire is actually asking for help? I never thought the day would dawn."

Sam smiled and smoothed his dress. "Not help, exactly. Let's just say that you have more extensive contacts." It was an understatement. Colton had exploited society's underside, making contacts behind the front of his import/export business to build a smuggling ring that spanned the globe. He prided himself on having people everywhere who could "acquire" anything. Drugs were boring to him. It was things of *real* value that he sought for his buyers— works of art, manuscripts, coins—all priceless originals and all stolen.

It didn't take very long for the Leaper to figure out he was there to nab Colton. But there was one point of which Ziggy had failed to inform him.

"And you would like me to provide you those contacts," Colton said. "My dear, have you forgotten the first rule of our business? I rather hope I have trained you better."

Better than you thought. When he started digging for evidence to get Colton, Sam found enough dirt to fill the Holland Tunnel. But he also uncovered a paper trail, so well hidden that not even Ziggy had suspected it, leading directly to Ann-Marie.

She was in it, as Al put it, up to her eyeballs. Sam's hindsight then allowed him to see the obvious evidence that surrounded him. The woman held a top position in a leading international firm. Her office was decorated with the effluvia of many cultures—tribal masks, hand-painted rice-paper screens, and the like—that Sam, with his late 1990s sensibilities, had failed to find odd for a woman in the male-dominated workforce of 1975. He figured it was just the trappings of the job.

He hated Al's initial she-slept-her-way-to-the-top theory. He hated even more that his friend was probably right. Women didn't generally move in power circles of this type in the mid-1970s unless they used every bargaining chip available to them.

Now Ziggy said he would have to find some way of getting Colton without destroying Ann-Marie's life in the process.

"Price first," continued Colton. "What do you offer me for my troubles?"

"Well, they're *your* contacts." Sam sat down on the edge of the desk, crossing his legs and arching his back. "I guess *you* can name your price."

"Really, Ann-Marie, seduction is so unbecoming, and my services don't come so cheap. Offer me something I can work with."

"Fifteen percent."

"Hah! Not bloody likely. Eighty."

"Too high, Sam. Ann-Marie would never go for it."

"Eighty? I'm doing all the work."

"Come, now. It's my people you're using. Without me, your deal is dead. Eighty. Take it or leave it."

"Forty."

"Sixty-five."

"Deal." Sam injected just the right amount of resignation in his sigh. "Now the names."

Colton took a notepad and pen from the desk. "There's a man, Laslow, in Germany." He wrote a number and handed the paper to Sam. "Tell him I sent you. He's been in the smuggling game as long as I have."

"Bingo, Sam!" said Al. "That did it."

"Now," continued Colton, "I think I *will* accept that other payment you offered."

Sam smiled and held out a hand. Colton bent to kiss it, and Sam's open palm met his nose in a bone-snapping crunch. Colton fell to the floor with a yell just as the door burst open and FBI agents swarmed into the room.

"Not bloody likely," Sam said and turned to the man approaching him. "I guess you got all you needed." Sam took off the wire that wrapped around to the transmitter on his back and handed them to the agent. It was the only solution left, really.

"We got it, but I still can't figure out why you came to us. The FBI has been watching Colton's operation for a few years, and not one of our agents ever found anything to prove you were involved."

"Well, Agent Barnes," said Sam, "I wanted out, and this way I don't become a loose end that has to be eliminated."

"I see what you mean." Barnes gestured toward Colton, who was snorting defiantly at the other agents through his shattered nose. "At least you went out with a bang."

"Is our deal still good?"

"Like I told you, you're in too deep for me to give you any guarantees. But I'll stand by you and push for the lightest sentence I can." Barnes smiled. "We couldn't have done it without you."

Sam held out his wrists. Barnes cuffed him, and he started to follow quietly.

"Colton goes away for a long time, Sam," Al said. He pushed some buttons. "Ziggy says she can't find anything on Ann-Marie. She's disappeared. Maybe she went into the witness protection program."

Sam felt a familiar tingle take hold. Whatever happened from this point on was moot. His job was done. He winked at Al.

And he Leaped.

CHAPTER ONE

Samantha J. Fuller sat hunched in the blue glow of the computer screen, working on the ideas that had popped into her head during dinner. She was close now, she could feel it. The equations marched across her monitor, an ever-changing army of numbers and symbols tramping toward their goal.

A tingle of excitement flowed from her stomach directly through her fingertips, which were flying over the keyboard like pheasants avoiding buckshot. She mentally reviewed her basic postulates as she typed, checking the formulas for accuracy. They were the crucial foundation of her work. They had to be right.

And they were, as far as she could tell. She entered the last digits and sat back, feeling relieved and not a little bit smug. "My dear Dr. Beckett," she said, "you're coming home." She regally poked the Return button.

An eternal moment passed, and then the monitor went dark. When it came back on, a hodgepodge of cybernonsense broke through her neat rows of logic, wreaking havoc without mercy. She barely had time to gasp before the screen flashed again. All that remained was a single sentence—three words that seemed to crouch around the screen's prompt in an attempt not to be read.

Does Not Compute.

"No, no . . . shit!" Samantha pounded the keys in frustration. "Ziggy!" she yelled, the Southern twang in her voice more pronounced with tiredness and frustration. "Where did I go wrong?" A new sentence started to type itself on the monitor and Samantha turned it off in disgust. "Talk to me, dammit!"

"Your attempted retrieval formula did not take into account the fluctuating variables of neuron and meson retention as they pertain to an individual Leap," came the unperturbed reply.

Goose bumps prickled Samantha's skin at the sound of the cool voice. She never seemed able to get used to the eerie tone of the parallel-hybrid computer's speech. The voice seemed to float out of nothingness and rivaled her Grandmama Laura's in the ability to cause the heebie-jeebies. "You also failed to realize that the odds of retrieval decrease exponentially, not linearly, with each successive Leap due to the constant merging with outside subjects."

"So you're saying the problem is scrambled brains," said Samantha, cutting the computer off. As frustrating as it was, it made sense. The momentary merging of the doctor's psyche with that of his host as they switched places led to the exchange of neurons and mesons. That was inevitable. Samantha could see why her equations fell short.

For someone who had been Leaping as long as Dr. Beckett had, the chances of recovery were slim indeed. His neurons and mesons had been swapped like baseball cards far too many times over the last five years for the computer to get a good enough fix on him to pull him back. But that led to other questions.

"If we can't get an accurate enough hold on Dr. Beckett to bring him back," she asked, "then how are we able to track him at all?" She already knew the answer, but wanted to make sure she had missed nothing.

"Three ways," Ziggy replied. "Foremost is the link that Dr. Beckett and I share. My neurocomponents were developed from his cells. As long as he remains alive, I will always be able to track him to a degree. Then, of course, there is Admiral Calavicci, who can achieve a neuron lock in the Imaging Chamber...."

"And last, but not least," Samantha finished, "we have the person in the Waiting Room. But won't this continual swapping eventually cause Dr. Beckett's consciousness to dissipate completely?"

"No, Dr. Fuller," Ziggy replied. "When Dr. Beckett Leaps out of a specific time and person, most of the neurons and mesons he lost are returned. For the amount of time he is in flux, he is whole."

"So what you're telling me is that the best opportunity to bring him back is when we have absolutely no way of knowing where he is!"

The facts didn't seem to bother Ziggy. "That is correct."

Samantha slapped her desktop unit off and stalked to the closet. She pulled out an old M.I.T. sweatshirt, drew it over her head, and made for the door. It was the proverbial catch-22.

She walked through the Project's corridors, trying to punch a hole in the problem, but it wore diamond armor. She put it in the back of her mind. Maybe her subconscious would work it out for her. Aside from the skeleton crew of officers stuck with pulling the night shift, Samantha had the halls to herself. She knew it had to be at least two in the morning. She decided to go to the cafeteria and see what goodies were to be had. Nothing helped her work through a problem as well as a pint of banana-fudge superchunk ice cream topped off with a cup of hot cocoa.

The holographic image surrounding Al swirled and abruptly disappeared, leaving him in the cold blue glow of the Imaging Chamber. It was his favorite sight. It meant that Sam was one Leap closer to home.

He stepped off the silver disk located at the room's center and fingered his hand link. The door slid up with a hydraulic sigh. He walked through the light and into the Control Room.

"Ziggy," he said to the blue orb suspended above him, "what *did* happen to Ann-Marie?"

"I am having difficulty accessing that data. Try back in eleven-point-seven hours."

"Eleven hours? Since when does anything take you eleven hours to figure out?"

"Since now." The tone was haughty, almost daring challenge. Al was in no mood for an argument, so he decided to drop it.

From the general quiet around him, he figured it was the wee hours of the morning. He wasn't the least bit tired. Leap-lag was one of the few dangers of Observing. He found his hours dictated predominantly by the time and place Sam had Leaped into. It was sometimes a bit disconcerting to walk directly out of a sunny afternoon into a deserted Control Room with the main console humming quietly on autopilot. He knew Ziggy would call Tina or Gooshie right away if anything went wrong.

What *really* worried Al was that Gooshie might see his absence as an opportunity to weasel his way into Tina's good graces. How could she be attracted to both of them? He was a former jet jock, one of the first men on the Space Shuttle and a national hero. Gooshie was just . . . Gooshie.

He took a puff on his cigar. "Ziggy?"

"Yes, Admiral?"

"Would you tell Tina to meet me in the cafeteria?"

"I am not a paging service, Admiral. But even if I were, Dr. Martinez-O'Farrell is sleeping at the moment." Ziggy paused. "If you'd really like to see her, I'm sure she'd be a lot more receptive if you . . . woke her *personally*."

"Get your mind out of the gutter, girl," Al shot back, a grin growing on his lips. That was one less worry. He bade the computer good night and went out of the Control Room toward the cafeteria.

When he got there, he was surprised to see Sammy-Jo Fuller polishing off what had once been a big bowl of ice cream. He had something less sweet, but no less sinful, in mind.

"Al," Samantha said, "I take it Dr. Beckett has Leaped?"

Al nodded and went behind the counter into a well-stocked kitchen. When Sam had laid out the complex's building plans, he'd paid almost as much attention to the kitchen's design specs as he had to the Accelerator Chamber's. It had made little sense to Al at the time, and he tried to dissuade his friend from being so lavish. It was just the kind of spending that would earn them top billing on one of the tabloid shows. He could just picture it: "Quantum Whiz Kid Turned Junk-Food Junkie—at The Taxpayers' Expense, on the next *Roberto!*" Every housewife in America would be out gunning for blood.

"Try not to see it so much as a budgeting nightmare as a morale booster," Sam had quipped. "Being buried under a mountain in the middle of nowhere is one of the job requirements of working on the Project. A sense of home will make even Weitzman happy."

And it had, for the most part. Sure, there were still knock 'em down-drag 'em outs when it came time for the Committee to review the annual budget, but they seemed perfunctory at this point. Weitzman initiated them, of course. But if the way he stuffed his face during the annual Committee visits (stovepipe hat tilted in a manner that clearly meant business) was any indication, he did it more for the pleasure of baiting Al than out of any real monetary concern. So Al continued to make stocking the kitchen a chief priority when he was wearing his Budget Manager hat.

He was very glad for it at the moment. He returned to the dining area carrying a prepackaged bean burrito and about a gallon of guacamole and hot sauce. "A little victory dinner," he said, putting the burrito into one of the compact microwaves located next to the coffeepots at a far end of the counter. "And now I don't have to go to heartburn city alone."

"So I take it Dr. Beckett straightened everything out for Ann-Marie Renerie?" asked Samantha, brushing her hair back. Al eased into a chair across from her, amazed at how much she looked like her mother.

When he had come out of the Imaging Chamber and seen her for the first time, standing behind the main console, it was all he could do to keep from yelling, "Abigail!" She seemed to be somewhere between amused and confused when she saw the look on his face. "What's the matter,

Admiral?'' she had asked. "You're acting like you never saw me before."

He had recovered quickly, of course. It took a lot to shake the Calavicci poker face, perfected by years of dealing with blood-sucking divorce lawyers. The newly forged past had come to him slowly, but he remembered it, right down to Sam introducing them on her first day on the Project. Time travel played mind games almost as well as the women he'd known.

Al shrugged. "I don't really know. I guess he did, because he Leaped. But Ziggy didn't have any answers for me. The only thing I can figure is that she plea-bargained her way into a new identity." He sucked on his cigar and quickly tamped it out when he caught the expression on Sammy-Jo's face. "It's out, it's out," he said, being careful not to exhale in her direction. "Don't wrinkle your nose at me like that, Sammy-Jo, it makes your face all scrunchy." He was the only person who called her that. She thought it was just his way of teasing her. He couldn't tell her the real reason—that had been the name he'd first known for her, when she was ten years old.

"Oh, that's *disgusting*," she said as he placed the stub in his vest pocket. "How can you walk around with that smelly thing? It looks like a slimy, half-eaten Tootsie Roll."

"I have my slimy vices, you have yours," Al said, gesturing toward her ice cream. Ever-growing pools of vanilla overtook lonely bits of banana.

"Oh, that." Samantha took her spoon and mashed the last of her snack into a yellow-white paste. "I'm just drowning my sorrows."

Al got up at the sound of the beeping microwave and juggled the hot food to his plate. "Want to talk about it?"

"Far be it for me to ruin your victory party," she answered. He remained silent in the hope she'd continue. She did. "I was *this* close." She held up her thumb and forefinger in exasperation. "This close to getting Dr. Beckett home."

Al stopped chewing, scalding beans forgotten for the moment. "What do you mean?" he asked, swallowing quickly. "How?"

"I had worked out a retrieval formula. It seemed so right. But it was another dead end."

Al searched for something to say. He knew that getting Sam home had become Sammy-Jo's pet project. But to his knowledge she had never done more than theorize. He had even let her use Sam's private office in the hope that it would strike some spark. Except for Donna, he had declared the room off-limits. It was Sam's fortress of solitude, in a way— the place where he did his thinking. It would be exactly as he'd left it when he returned. But considering the bond Sam and Sammy-Jo shared, even if they didn't know about it, Al had granted her access—maybe the room's aura or whatever would kindle some dormant inspiration in the woman. Fact or fancy, Al didn't care. When it came to bringing Sam home, anything was worth a shot.

"Don't take it too hard," Al finally replied. "Not even *Sam* can get Sam home, and he's the wunderkind."

"That doesn't matter, Al. I'm a genius too, remember." It was not a brag, Al knew, only a simple statement of fact. "And I kinda feel like I owe it to Dr. Beckett to get him home."

"Owe him?"

"Before I met Dr. Beckett, I was floundering. I was done with M.I.T. and on my way to work at a computer firm in Mobile, of all places. Then one of my professors, Dr. Lo-Nigro, introduced us.

"I know it's going to sound crazy, but I felt a bond at that moment. I never told you this, but I grew up without a father. That always left sort of a hole. When I met Dr. Beckett, that feeling disappeared. Another person made it go away once before that." Sammy-Jo's eyes took on a faraway look, as if straining to see something shrouded in the cloak of years. "I was only a little girl then."

Al suppressed a grin. He would bet his next round of alimony payments that the "other person" was someone Sammy-Jo Fuller had known as Sam Larry Stanton. "Well, you still shouldn't beat yourself up, regardless. Not for failing at the impossible." He didn't know what else to say. *The girl may look like her mother, but she's her father's daughter. She wants to save the world, too.*

"Cheer up, Al," she said, brightening. "I told you I wouldn't ruin your victory party, and I won't." The tone sounded forced to Al. He knew of a sure way to make her smile again; he could tell her the truth, Project rules be damned. He knew the pain of growing up without parents, and he was loath to see anybody else go through it.

But that would probably lead to a fiercer determination on her part to do the impossible and to further self-abuse when she realized she couldn't.

"If anything," he said, "it's my fault he's still Leaping around. He was home, remember, and gave it up to save me." It was a calculated risk. The swiss-cheese factor had made the events surrounding his jaunt to the past a bit sketchy. All of it had happened "before" Sammy-Jo Fuller, Sam's daughter, was part of the Project. But he supposed that once she was introduced to the time line, the original history had reshaped itself around her and incorporated her into the event.

"Don't be silly, Al," she replied. "You can't possibly believe you're to blame. Dr. Beckett was acting of his own free will. You had no control over that."

He was glad she *did* remember it; he would have felt foolish otherwise. "And you *do*? Did *you* push Sam into the Accelerator Chamber?" He had trapped her in her own logic, and she knew it. He pressed his advantage and answered for her. "No. He did it of his own free will. That's nobody's fault, least of all yours." He punctuated the last point with his burrito. "So I don't care what kind of bond you may feel. It still doesn't make it your responsibility."

"Brilliant coup, Admiral." Sammy-Jo rose, snapping a mock salute. "Permission to go to sleep, sir?"

"Aye. And watch out for the proverbial bedbugs, or mountain weevils, or whatever the hell they are around here."

Samantha turned to leave, but paused at the door. "Thanks Al," she said, a soft grin framing the words. "G'night."

Al, returning her smile, watched her go. He lit his cigar and took a few long puffs. "Eat your heart out, Verbeena."

Long after Al had disposed of the dirty plates and retired for the night, Ziggy was still working. She had long since left the minor systems on automatic and focused all of her neurochips on locating the subject *Ann-Marie Renerie*.

New York Times Records library.
No data.
Federal Bureau of Investigation files.
No data.
United States Witness Protection Program.
No data.
Internal Revenue Service records.
No data.

The parallel-hybrid computer searched all the available files that might have information on the woman and some that in all probability didn't. She even went through the rather humbling process of repeating the procedure. It still yielded no information. Methodically, Ziggy searched the memory banks of all the prison computers across the country.

Ashland Federal Corrections Institution.
No information.
Danbury FCI.
No data.
Bedford Hills Correctional Facility.
Ann-Marie Renerie. Inmate number 064-E-76. Admitted 3-18-76, released 7-12/7-12/7-12/7-12/7-12 000000000000000 0000000000000000000000000000000000000000000000000 0000000000000000000000000000000000000000000000000 0000000000000000000000000000000000000000000000000 00000000000000000000000000000cfb-pd X/.*. 9upe—
No data.

Ziggy searched the records again. It was fruitless. The information that had come up before was no longer there. There had to be a logical explanation. Perhaps it was in her own circuits.

She ran a standard diagnostic program. The results remained in the optimal percentage range. Something to do with the other computer? Possibly. She checked for the data again and again, but to no avail.

The parallel-hybrid computer continued searching. But as the night wore on, an alien feeling encroached on the cool logic of Ziggy's neurocircuits. Under ordinary circumstances, the only "feeling" the computer could be said to have was one of superiority. Now, for the first time ever, that superiority was wavering.

CHAPTER TWO

The bell's harsh clang echoed hollowly off the concrete walls, abruptly rousing Ann-Marie out of her doze. The copy of *Town & Country* slid from her chest as she sat up, society pages folding to the floor and shutting her out.

"How poetic," she murmured, rubbing her eyes. She kicked the magazine, and it landed between the sink and toilet. The latest fashions from Paris stared back at her.

She ignored them and turned on the faucet, hoping for warm water to wash with. The faucet's noise followed her as she paced her cell—three steps, turn, three steps—following the faint impression worn into the concrete floor from the door to the stainless steel sink by countless feet taking the same route for countless years. She was carrying on a great prison tradition, it seemed.

Ann-Marie tested the water and splashed it on her face. Tepid was probably the best she'd get today. She supposed she should be happy. Some days, there was no hot water at all. Others, it was so scalding that you couldn't touch it. She grabbed the washcloth she kept folded on the corner of the sink and jumped back with a yelp as a cockroach fell out of it and into the basin.

Of all the things she had had to get used to since coming to prison, roaches were the worst. Life on the inside took a lot of mental strength, and she'd found a good source for that so far. But one night early on, she had awakened when she felt a light tickling all over her body. She opened her eyes to see her arm covered in a writhing brown mass. The scream that welled up died in her throat when she felt something moving on her tongue. One of the vile things had decided to forage in her *mouth*. She spit it out and vomited. *Then* she screamed; a scream to wake the dead. There were hundreds of them all over her and the bed.

Tibor, the bastard, actually laughed when he finally opened the door and saw her flailing around the cell with roaches flying from every part of her. For an encore, she slipped in her own puke and fell to the floor in tears of rage.

"Get up," the CO said, still guffawing. She kicked him in the crotch instead. It had the desired effect; he took her to the hole and she stayed there for three days, relatively roachfree. The charge sheet cited her for assaulting a Corrections Officer and causing a disturbance. Nowhere did it mention that she was sleeping on a mattress that had turned into an insects' nest.

She definitely couldn't take roaches. She crushed the one still struggling to get above the waterline and watched it swirl down the drain.

If she had ever seen a more concrete symbol for the turn her life had taken, she couldn't remember it. She tasted bile in her throat as anger flared in her. Even with all the order prison imposed on her life, none of it made sense. *How the hell did I wind up in this godforsaken place?*

That was the $64,000 question to which she had no answer. She had been on top of a million-dollar operation with her ass covered so well that even she couldn't find it with both hands and the proverbial flashlight. Then, one day about three years ago—*Have I been in this place that long?*—it all fell apart. No explanation. No rhyme or reason. They had just taken it all away from her. And try as she might, she still couldn't figure out how it happened.

Colton had been her first guess. Toward the end, she really didn't need him anymore, and he knew it. They both knew it would come to that one day—well, she had, anyway—the classic story of the student surpassing the teacher. Only she had anticipated it and was prepared. The evidence she had squirreled away with specific instructions on who to give it to in the event of her untimely demise (say an ice pick in the ear) had kept the bastard in check, and should have for good.

But even when it seemed apparent that it hadn't (not surprising, due to the man's childish mentality and jealous tendencies), she was still prepared. No one would believe that a woman could climb so high in the old boys' club of big business, robbery, and corruption.

Especially not a jury.

It would just be another case of male domination and a woman too confused with love and hero worship to know what she was really doing. She was prepared to play Pitiful Pearl if it suited her needs.

What she wasn't prepared for was the truth.

There was still no way she could believe that she had done this to herself. Tape recording or no, signed confession or no, it was impossible. It couldn't happen. But, apparently, it had. And try as she might, she couldn't divine an answer. The more she tried to remember what had happened, the more frustrated she became.

It was as if she had fallen asleep while getting ready for work one morning and awakened in handcuffs. Between the two events was a chasm so vast and final that it never yielded the answers she knew lay at the bottom. And three years later, no matter how deep she delved, Ann-Marie knew that she was still barely skimming its surface.

She took small comfort in the fact that she would have seven more years of leisure time to dwell on it. The thought made her shudder. As much as she tried to tell herself that seven years wasn't that much longer, 1986 still seemed more of a vague notion than an actual point in time that she could look forward to. How much would she miss?

She quickly finished washing up, trying to erase the thoughts from her mind. But as she turned off the faucet, the magazine crumpled in the corner caught her eye. The rage returned and she scooped it up, balling loose pages in her fist. *There has to be some way to explain this! There has to be some way I can get even!* She took a few calming breaths and smoothed out the magazine to prove that she was still in control. If Tibor saw it crumpled on the floor, he would probably write her up for littering.

She walked out of her cell, joining the other women lining up for dinner. Tibor and Hernandez, another CO, sauntered up and down the corridor.

"Let's get in line, ladies," Tibor yelled. "Fall in."

"What's it look like we're doin'?" someone asked, just loud enough to be heard over the din of shuffling feet and muttered curses that seemed to comprise most of the prisoners' vocabularies. It was Evangelene, Ann-Marie saw, and she got in line next to her neighbor. "We been doin' this every day for how many years now?" Evangelene asked, turning to Ann-Marie. "Why do they have to tell us what to do if we're already doin' it?" Anger sharpened her Southern accent, making Ann-Marie smile.

"It makes them feel useful."

Evangelene snorted. "When've they been useful for anythin'? 'Cept maybe to cause misery. A gentleman would never act like that. My *daddy* would never act like that. He'd teach 'em some manners."

"Enough, Angie," Hernandez said from behind them. "Fall in, now." He started to walk on, but Evangelene cleared her throat loudly, stopping him.

"Excuse me, Mr. Hernandez," she said as he turned back, "but I feel I must remind you of somethin'. God took great care to properly name all of his creatures. And my daddy, being a God-fearing man, took great care to properly name all of his children."

"Good for him, Angie," Hernandez interrupted. "Now please fall in."

"*Excuse* me," Evangelene rode over him, her voice taking on the quality of impeccably mannered steel. The look in her eyes made the stone walls seem soft and rooted Hernandez where he stood, his words forgotten. "As I was sayin', my daddy took great care to properly name all of his children. He named my brothers Tobias, James, and Cody. They were referred to as Tobias, James and Cody. My sisters were christened Ernestine, Isabella, and Theodora, and that's what we called 'em. He named me *Evangelene*." The slight woman seemed to tower over the broad man with the sheer conviction of her words. "Not Angie. Not Lena. Not Eva. *Evangelene*. I would appreciate it if that's how you referred to me. Considerin' the fact that I'm old enough to be your grandmother, you *should* call me ma'am." She held up her hands in a surrendering gesture. "But I know that I'm not your grandmother and you have your authority." The word seemed mocking, even though Evangelene didn't emphasize it in any way. "So Evangelene will do."

The guard stood in stunned silence for a beat until he realized the prisoner was done. "Yes, ma'am," he said quickly under his breath, seeming to surprise himself. Anger sprang up on his face and he opened his mouth to say something, but he closed it immediately, realizing, Ann-Marie guessed, that anything he said now would only make him look even more foolish. He scowled at his shoes and turned away, walking up the line. "Fall in!" he practically bellowed. "Let's go, ladies!"

Evangelene leered at his back, falling into line next to Ann-Marie. "Like kindergarten, it is. My Thomas, may God watch over his soul"—she made the sign of the cross absently, anguish briefly painting her features—"had more manners at ten."

Ann-Marie chuckled under her breath. Admiration didn't begin to describe what she felt for Evangelene. The COs might be in charge, but the older woman carried herself with a quiet (and sometimes not-so-quiet) dignity that implied she would have no problem sending anyone who stepped out of line to go cut a switch and meet her behind the woodshed.

The woman was a beacon, a Southern belle with no ball to attend, whose sane light had guided Ann-Marie through the insanity of prison life from day one. At first, it was the only thing that pulled her through.

Her first day had been a humbling experience, to say the least: faceless gray-clad guards leading her into the mammoth facility in chains. It was even uglier on the inside, smelling of damp and mold.

She and the other new prisoners awaiting processing were packed like cigarettes into a holding area. She kept her eyes to the floor. Her pride wouldn't allow her to show the other women the fear and shame that were overcoming her anger.

The officer waiting with them blew her whistle, told them to get back in line, and began removing the chains. A bell rang in the tiny room, rattling Ann-Marie's nerves. The guard herded them through the door on the other side of the room and Ann-Marie took the first step into what seemed like a downward spiral to hell.

Every step of the processing—from the showering to the medical examination to the cataloging and storage of the last personal possessions she had managed to keep with her— stripped her of her last vestiges of cocky self-assurance. By the end of it, she was dressed in a pale green jumper exactly the color of the scrubs her father used to wear while working in the hospital, the number 064-E-76 stamped above her left breast. She was no longer Ann-Marie Renerie; she was the sixty-fourth inmate brought to E-Block in 1976.

Another female guard led her through more musty corridors until they came to a black-painted iron door with a wire-reinforced window set into its upper half. The officer tapped on it with the end of her nightstick. A guard on the other side looked through the glass and opened the door, causing yet another bell to clang. This one pierced Ann-Marie's soul.

"New tenant," the officer said to her colleague as she led Ann-Marie into the block. The stout woman behind the desk nodded and picked up a clipboard.

"Ann-Marie Renerie?"

"Yes," Ann-Marie replied under her breath. The shaky voice seemed to come from somewhere else.

"I'm Corrections Officer Ward. Are you listening?"

Ann-Marie nodded, looking the other woman in the eye.

"My name is Ward," the CO repeated. "I'm in charge of this block during the day, five days a week. If you have any problems, tell another officer and they will tell me." Ward stood, bringing her bulk around to the front of her desk, coming toe to toe with Ann-Marie. "I don't anticipate any problems. The rules are posted at each end of the block"— she gestured to a yellow sheet of paper stuck to the wall next to the door—"and they are to be followed to the letter. Any new rules will be posted by memo. It's your responsibility to keep up with them. Lockdown is twenty-three hours a day, with one hour of exercise in the afternoon. Wake-up is at six, lights-out at eleven. Between meals you can stay in your cell or get to know your neighbors on the block. It's your choice. You will do what the other officers tell you to at all times. If not, you will be written up or put into solitary, depending on how serious your offense is. Any questions?"

The tone of Ward's voice said she clearly didn't expect any. Ann-Marie shook her head. "Good," Ward continued, resuming her place behind the desk and waving another officer over. "Tibor will escort you to your cell."

Ann-Marie turned to the guard approaching her—Tibor, she supposed—and began following. Women were scattered along the corridor in small groups, staring at her with dead eyes and whispering to one another. The talking got louder after she passed, the women getting braver when they didn't have to face her. More seemed inclined to joke with the guard at her side.

"Got yourself a new girlfriend, Tibor?" said one woman who walked out in front of them, shaking her curly black hair back and forth. "I thought I was your girl."

"Move it, Rosa," Tibor replied. "You know I only have eyes for you." Rosa darted into an open cell on the other side of the corridor and peered at Ann-Marie from behind the door frame, sizing her up. Ann-Marie quickly turned her head away and noticed that every woman ahead was looking at her the same way. She was on display—a new sow being brought to auction. She turned her eyes to the floor, away from the degrading stares.

"What's the matter, girl?" yelled one of the women up ahead. "You think you're too good to be looking at me? You're no better. We both got numbers. You're a number the same as me." Other prisoners began parroting the first one, yells asking who she thought she was and did she think she was better than them and was she too good to be here with them? Ann-Marie felt heat rising in her cheeks at the continued hoots and demands as they rose to a cacophony that shot through her and caused something to snap.

"Yes!" she yelled in the rage that had awakened in her, silencing the others. "Yes, I *am* better than you! Better than every damn one of you. I don't belong here! I don't belong here!" She turned and started walking the opposite way down the corridor. Tibor was on her in an instant, grabbing her by the shoulder and trying to turn her around.

"This way," was all he said. She shrugged him off and kept walking. "Take your hands off me!" she spat. "I'm not one of these animals."

Tibor grabbed her around the middle, pinning her arms to her sides, and lifted her, struggling to keep her still. She fought back with everything she had. Other guards joined Tibor and surrounded her, pinning her to the floor and cuffing her hands behind her back.

"Get off me!" she screamed, her voice getting lost in the din now coming from the other women. "Don't you touch me! Any of you!" Four guards, two on her shoulders and two on her legs, lifted her off the cold concrete and carried her struggling form the rest of the way to her cell, where they dropped her on her bunk and locked her in.

Ward's face appeared at the grate set in her cell door, her eyes as calm and steely as her voice. "Done?" It wasn't really a question. "Congratulations. You just earned your first charge sheet. Creating a disturbance, assaulting an officer, and disobeying rules. Any one of them would earn you a few days in solitary, if not worse." Ann-Marie couldn't imagine anything worse. "But I'm feeling generous. Considering it's your first offense, I'll go light on you. No dinner tonight."

Ward turned to leave, but stopped short and pinned Ann-Marie with a gaze that made the previous one seem friendly. "Plus I want you to think about what you did. And I don't mean that in the way a parent or a teacher would. When I say 'think,' I mean understand something. *Really* understand. You no longer have the power that you did on the outside. You are subject to the rules of this facility and have absolutely no say in the matter—not now, not in the future. Forget the B-movie notions of favoritism and earning a better place for yourself. You don't do me favors and I don't do you any. There are no queens on my cell block, only those who follow the rules and those who don't." Ward continued staring when she was done talking, as if trying to drive the words into Ann-Marie's marrow. "I trust there won't be any more theatrics. I will unlock your cell after the other inmates have eaten, so you can meet them. Although I don't know if you'll want to come out after what you called them." Ward left Ann-Marie to stare at the section of empty corridor that she could see through her door's window.

During the conversation, if it could be called that, her rage had dissipated like fog facing the sunrise, leaving tears in its wake that made hot paths down Ann-Marie's cooling cheeks. She buried her face in her hands and let the sobs come.

She stayed that way until a knock on her door brought her back to her surroundings. A small black woman stood in the doorway, half in and half out of her cell.

"May I come in?" the woman said, stepping into the cell. "My name is Evangelene Robbins. We're neighbors." She looked at Ann-Marie and scowled. "Wipe your nose, girl."

Ann-Marie wiped her nose with the back of her hand.

"That's better," Evangelene said. "It's not proper for a lady to go around lookin' a mess."

"What do you want?"

"I make it a habit not to speak to strangers. And seeing that you haven't properly introduced yourself, we're still strangers."

"My name's Ann-Marie Renerie." *Who the hell did this woman think she was?* "Are you for real?"

"Considerin' the fact that I'm standin' here, that's a very silly question."

Ann-Marie almost threw her out in annoyance, but Evangelene intrigued her just enough that she didn't. Maybe it was the brightness in her eyes, which belied her advanced years. She didn't have the hollow stare of the other women Ann-Marie had seen earlier.

Another bell rang, snapping Ann-Marie out of her reverie. The dinner line started moving toward the iron door at the end of the cell block.

Befriending Evangelene was the best thing she had ever done. She had taught Ann-Marie the key to prison survival, the key to real power both inside its walls and out: dignity. The belief that Evangelene had in herself enabled her to carry herself with a boastless pride that belittled her circumstances.

"ID," said the guard from her station at the door. Too late, Ann-Marie realized she had left it in her cell. *Those damn roaches*. They halted the line for her and an angry buzz rose from the women still waiting. Most times she would be forced to skip dinner, just to teach her a lesson. But Tibor must have been in a good mood. He escorted her back personally.

A blinding light tore through her and hurtled her soul into a blue void that expanded forever. She flailed through it at an impossible speed, yet could feel no sense of movement, no taste, no smell.

Horror ripped through her, a dread that had no equal. She was lost, cast adrift. Alone.

A sudden force pulled her, drawing her atom by atom into the center of a swirling vortex that raged eternally.

The nebula coalesced around her, gaining substance. Feeling was returning. She could faintly sense a heart beating, blood circulating. Cool air brushed against her skin. Sound met her ears, and she turned her head toward it.

Vague shadows united and became two people wearing white coats, standing within arm's reach and speaking softly. The feeling of horror that had engulfed her before increased tenfold, and she let out a scream—

Ann-Marie sat up with a start, gasping in the darkness that surrounded her. Clutched by fear, she reached over and switched on the small lamp on the nightstand next to her bunk. The light threw odd shadows in the familiar confines of her cell, low wraiths that she found more disquieting than the dark. In three years she had moved to better quarters, behaved herself, gotten privileges. But she was still in prison.

She went to the sink, splashed cold water on her face and rubbed it through her sweaty hair. The nightmare images faded quickly, leaving only a vague feeling of unease. She looked into the mirror and forced herself to take a few deep breaths, willing her features to calmness.

Nightmares had been a fairly regular occurrence ever since

she got to prison. But she was sure these dreams were about more than having her freedom stripped away. They went deeper, somehow, and were coming more often. She could never remember them afterward. The only calling card they left was cold sweat and confusion.

She was sure they had something to do with the lapse of memory she suffered, but she couldn't figure out how or why.

Ann-Marie stripped off her drenched T-shirt and pulled a fresh one over her head. She sat on her bunk fingering the law books stacked neatly on the night table, books that she hoped would show her some loophole to freedom.

The rest of the table was covered with folded figures, paper animals for her own private zoo. At first she had taken up origami to help pass the time, but the delicate creatures had become something more; the skill necessary to create them was tangible evidence that she wasn't just an animal living in a cage. It was one of the little tricks that Evangelene had taught her.

But they offered little solace now. The perpetual wondering took its toll and demanded attention now and again, no matter what mental discipline she tried to maintain.

How? The thought came upon her so quickly and furiously that she swept the meticulously folded figures onto the floor and was halfway across the cell before she realized it. How did she wind up here? The answerless question weighed her soul down into the now familiar terrain of blind wonder.

I can't stand by and let it strip me of my sanity! Someone did this to me. I'll figure out how and who, and they'll pay for it.

She sat down again and took some sheets of paper from the bed table. She began folding, fighting to return to calm, absentmindedly switching the radio on low and letting the announcer's measured tones fill the silence.

Despite Ward's little speech, there was a hierarchy on the block; everyone but Ward seemed to know about it and it didn't take Ann-Marie long to move into its higher circles. Hence the radio and other little bits of civilization.

The plight of the Americans held hostage in Iran led the newscast. Other events of import were also reported, and the broadcast ended with a kicker about boy genius Sam Beckett being scheduled to give a piano recital in Carnegie Hall at the tender age of twenty-two.

Ann-Marie heard none of it. She was too busy concentrating on what kind of plans she could lay to find out who had wronged her. It would take time, she realized, considering that she had nothing to work with. But she had plenty of time.

CHAPTER THREE

The blinding blue maelstrom died as quickly as it had come, and feeling returned.

Cold.

It was the only thought that went through Sam Beckett's mind before he realized he couldn't breathe. Blackness surrounded him, and the panic that arose was blacker still. He flailed his arms and attempted to raise his head, but something was holding him face down in a murky soup that filled his ears and nose.

I'm gonna die! A muffled disco beat pounded through the ooze. *And this must be Hell's waiting room.*

Something slid under his face and hauled him up by the nostrils. He took a ragged breath and opened his eyes, but his vision was blocked by the hand tearing at his nose.

Martial arts training and reflexes took over and Sam grabbed his assailant's arm, twisting instinctively. The grip was slippery, but the attacker rolled off Sam's back and fell to the side, arm pinned.

Wiping his eyes with his free hand did little to clear his

vision, but he could see a crowd lit by blinking neon lights. They seemed to be cheering something. He ignored them in favor of getting a better hold on his attacker, but stopped short, staring.

His opponent was a woman covered in mud. The remains of a bikini top were wrapped around her stomach, leaving her breasts exposed. The shock caused him to lose his grip. She grabbed him by the ears and brought his face down into her bent knee. The crowd roared louder.

Sam landed on his back. He caught a glimpse of himself in the mirror suspended from the ceiling. A woman stared back, so covered in mud that he couldn't tell if she—*he*— was also topless.

A knee came flying out of nowhere and landed in his stomach, making him curl up. He fell flat again, his wind gone, as the other woman pinned his shoulders.

"One . . . two . . . *three!*" came a voice. "We have a winner!" A bell started dinging and the woman stood up, stomping around the ring with her arms raised in triumph. "Let's hear it for Tawny!" The announcer stretched every word, milking the crowd.

The crowd began chanting. "Taw-ny! Taw-ny! Taw-ny!"

Sam plopped his head back into the mud, groaning.

"Oh, boy."

He rolled to his knees and shook his head. When he stood up, a few drunken boos disrupted the mantra, causing the crowd to ripple with laughter. A rowdy bunch of men surrounded him, excitement and naked lust painting their features.

"And let's hear it for the opponent," the announcer continued, "Miss Candy Apple!" Applause rolled over the booing and Sam remembered that he was a woman, naked from the waist up. He quickly threw an arm across his chest, hoping that it would provide cover. It did, if the crowd's rising rumble was any indication.

He wanted to get the hell out of there, but had no idea where to go. He stepped out of the mud and joined Tawny and the announcer, hoping to take a cue from them.

"Congratulations, Tawny," the announcer said. "What does this victory mean to you?"

Tawny grabbed the microphone. "It proves that no one can stop me!" she yelled, to the delight of the men surrounding them. "Mark my words, Richard! I will be Mud Queen!" The whistles and cheering rose to a deafening level.

"And you, Candy? Will this stop *you*?" The microphone was suddenly under Sam's nose, shocking him out of his study of the crowd.

"Huh?" The announcer's eyes widened with insistence. "Oh. N-no . . . Richard. It will take more than this to stop me from becoming Mud Queen . . . yooouu betcha."

Tawny glared at him, then grabbed the microphone again. "We'll see about that!" she yelled, turning to the crowd. "I will reign! I will conquer! I will. . . ." She gasped abruptly, eyes going wide. Handing the microphone back to Richard and covering her breasts with her free arm, she quickly stalked off. Sam followed her, trying to ignore the hoots and whistles that trailed him out of the room. He could feel the blood rising to his ears. *What would my mother say?*

Tawny led him through a door and down a short hall. The mud was starting to dry on his face, creating an uncomfortable mask that pinched if he blinked. He was about to ask Tawny what happened, but she stopped at another door and turned on him before he had the chance.

"What the hell was that?" she said, the anger in her eyes the only expression visible through her own mud mask. "You almost broke my arm out there!"

"And you almost drowned me!" Sam shot back. "What was I supposed to do?"

Tawny shook her head, mud flaking like ashes to reveal the tips of her brown hair. "*Let* me. You knew I'd pull you up. That's one of the oldest moves in the game." She opened the door and walked in.

Sam followed her into a small room. He didn't think it could be called a dressing room. It looked more like a lounge. More women crowded inside, all wrapped in towels, hair dripping, with the exception of two in clean bikinis who were sitting on one of the low couches that lined the velvet-papered walls. Cigarette smoke hung thick in the air.

He was glad to see the two bathrooms at either side of the room. At least he wouldn't have to wait for a shower. He was less glad to see the Observer standing in the corner.

Al's grin nearly split his face in half, and his eyes gleamed as if he had just discovered Truth—and the beauty that lay within. His cigar stuck straight out of his face, forgotten, the smoke it created seeming to mingle with that of the girls' cigarettes.

Sam sighed and dug through the bags piled next to one of the couches, hoping for some clue to tell him which to take. It wasn't hard; the bag with the bright red apple on a stick painted on its black canvas side had to be it. He shouldered it and started for a bathroom.

"You're on," Tawny said to the two girls waiting on the couch. They got up and walked toward the door, dragging a sigh from the Observer.

"So long, Kiki. Take care, Cookie," he said. He glanced at the bathroom Sam had started for. "See you *after* the match." Sam cleared his throat loudly. He cleared it twice more before Al noticed.

"Sam!" he said, clearly surprised. "How . . . ah, how long you been here?" He let out a sheepish chuckle. "I didn't recognize you under all that mud." Sam walked past him and shut the bathroom door.

By the time Al passed through the wall to continue the conversation, Sam was washing the dried clay off his face. "You're a pig."

"At least I wasn't rolling around in the mud," Al replied defensively. "Although I'd like to." He punched his hand link absently.

Sam turned the shower on. He paused. "Turn around, Al," he said sternly. The Observer sighed dramatically and turned his back. Sam extricated himself from what was left of his bikini. "I don't care what Ziggy says," he said as he stepped into the shower, pulling the curtain shut. "I don't want to hear that I'm here so that Candy Apple can become the Mud Queen." He let the hot water roll over him, removing the mud's chill.

"Candy Apple," Al snorted. "Hell of a stage name. Your real name is Sarah Bullock." He paused, whacking the hand link. "Ziggy doesn't really know why you're here yet."

Sam could barely hear him over the running water. "Does

32

she have any theories at least?" He didn't want to stay any longer than he had to. He'd rather be shackled to Diane Frost again. *Who the hell is Diane Frost?* Swiss-cheesing could be a pain in the ass. It was like remembering snippets of home movies about people he didn't know.

But there was one thing he always remembered hating about being a woman, and he knew that if he didn't Leap soon, the high heels would be calling. He finished washing quickly, anxious to get on with things.

"Can you at least tell me where I am?"

"Looks like hoo-ha heaven to me." Sam didn't find it funny. "Wilson, Arkansas," Al hurriedly continued. "It's a small city about twenty-five miles south of Little Rock. You're obviously a mud wrestler, and you work in what appears to be a small club, probably on the outskirts of town. It's October 25, 1988." Al paused, a smile spreading on his lips. "Hey! You get your Nobel Prize in about a month."

Sam glared at Al's back as he toweled off. *"And...."*

"And that's all Ziggy can give me right now."

"Come on, Al," Sam said, admiring Sarah's features in the mirror and combing back long blond hair. "I could have found this out for myself." He slipped into a bathrobe he'd taken from Candy's—Sarah's—bag.

"So you could."

"That's all you have to say to me? I build the world's most advanced computer and all it can give me is the date? You can't even give me the name of the club I'm in? What's wrong with her this time?" He rode over the beginning of the Observer's answer. "And don't give me some bull about a screwy chip from Pago-Pago." He paused. "If you promise not to leer, you can turn around now."

Al turned back to Sam, puffing his cigar. "We don't know." Sam remained silent. "She went a little blooie after your last Leap."

"Define 'a little blooie.'"

"She seems to be having trouble getting information. She still hasn't told me what happened to the person you just Leaped out of."

"How could that be?" Sam asked. "If I Leaped, I had to change history. Finding that change should be the easy part."

He rubbed the towel through his hair and twirled it into a turban.

Al studied the hand link. "Theoretically, that *should* be the easy part. But there are times when certain factors can screw it up. The Witness Protection Program is one of them. Other people just disappear."

Al was right, of course. In most cases, Sam heard the happy ending he had wrought. It was the thing he liked best. It almost made being stranded in time worth it. But at other times the Leap out was his only epilogue, saying that he did right or that he could do no more. He didn't like it, but that's the way it was. "Then if she can't find this person, why doesn't she just forget it?"

"Well, normally she would," Al replied, "but she's telling Gooshie and Tina that she can't this time. She keeps saying something about conflicting information, but she won't explain."

"Conflicting information is impossible," Sam said. "How can you get two histories from the same person? You have either the original history or what happens after I change things. Either way, events should be set."

"I don't know what's causing it," Al said. "Gooshie says it might be. . . ."

"Alia," Sam said softly. He might not remember much from Leap to Leap, but he would never forget her.

"No, no," Al said. "That's the first thing I thought of, too. Ziggy isn't reading another Leaper." Sam let out a breath he didn't know he had been holding. He didn't like the problem Ziggy was having, but he was glad Alia wasn't the answer. Killing someone was something he considered his last possible option in any situation. But he wondered if he hadn't made a mistake there. When he let Alia live, he knew he was defeating himself. But he just wasn't able to pull the trigger. The best he could do was hope that whatever had been Leaping her had dragged her back to hell and done the job for him. Not a nice thought, but better than—

"Sam." Al's voice snapped Sam from his musings. "Maybe you better get that."

"Huh?"

"The door." There was a knocking as Al spoke. "Maybe you should get it."

Sam opened the door to reveal a woman caked in mud, barely recognizable as one of girls who had left when he and Tawny returned. She had managed to keep her top on, but not by much. "Are you done in here, Candy?" she asked.

"Sure." Sam grabbed his bag and stepped aside, letting her in. "It's all yours."

"Cookie," Al's voice was filled with concern. "What did they do to you? Sam, it would be criminal if I didn't stay and help her clean up."

Sam glared at Al until he punched the hand link with a sigh and vanished. He smiled at Cookie and closed the door behind him.

A man approached Sam in the outer room. It was Richard, the announcer, with a handful of bills. "Quite a show," Richard said softly, stopping in front of Sam. The vocal animation he displayed for the crowd was almost completely gone, seemingly unsustainable without a microphone to prop it up. He counted out four fifties from his money roll. "I especially liked the arm thing." He looked around the room furtively, lowering his voice even further. "But don't tell Tawny I said that. She'd skin me." He pressed the money into Sam's palm.

"Thanks," Sam said, pulling his robe closed. "Don't worry, I won't say a word." Richard winked and moved on to another girl.

"Who's the *weeshp*?" Al asked.

"Isn't that what you're supposed to tell me?"

The hand link beeped and the Observer squinted at its tiny screen. "Ziggy's finally getting something. That's Richard Danson. He owns this club—The Kit Kat Bar—with his wife, Kathy Scherber-Danson." Al paused, frowning. "Why does that name sound familiar?"

"Maybe if you read on?"

"Quit being so impatient"—Al's concentration returned to the screen—"it's not ladylike. The bar is home of the Mud Queens, mud-wrestling troupe extraordinaire, which Richard helps his wife manage." Al gestured around the room with his cigar. "You're apparently just one of the many fringe benefits of his job."

"Don't tell me, Al," Sam murmured to the hologram. "I'm here to stop him from doing something to one of these women."

"Ziggy gives that only nine percent," Al said, much to Sam's relief. "Apparently, he's one of the quiet ones you *don't* have to watch out for. He treats them all very well. He makes sure they're paid on time, have clean bathrooms and a generally pleasant work environment. No need to worry about him; Dicky's a pretty stand-up guy."

The Observer paused again, trying to figure something out. "Richard Danson. . . . That sounds familiar, too."

"I'm sure the answer will come to you when you're in an environment with fewer . . . *distractions*," said Sam. He dug through his bag, pulled out a pair of jeans, and slipped them on. He discreetly drew a T-shirt over his head. He had perfected the art of getting changed in front of people without actually undressing; something he found necessary from Leaping into so many women. Not wearing underwear made it a bit easier. He had discovered early on that sliding into a pair of ladies' panties and a bra wasn't something he enjoyed. Walking around without underwear wasn't either, but it seemed the lesser of two perversions. "What am I doing here, then?"

The Observer frowned. "Give it a minute."

Sam spotted Tawny across the room, dressing in leisurely fashion. She seemed to have no reservations about baring all in a crowded room. Still, Sam waited until she pulled a T-shirt over her head before approaching her.

Tawny pinned him with an angry stare, and Sam raised his hands in a gesture of surrender. "I'm sorry about before," he said quickly. "I didn't mean to hurt you."

Tawny's ire dissipated. "Don't worry about it," she said, sighing. "I'm not really mad at you."

"Was it someone you saw in the crowd?"

Tawny looked up sharply. "What makes you think I saw anyone?"

"Why else would you run out of the room like that?"

"Look," she said, putting her money in the back pocket

of her jeans and digging her car keys out of her bag, "it's none of your business. I'm going home."

"Oh, no, you don't," came a woman's voice from behind them. "You still have to sign posters at the bar for that bachelor party. The groom is waiting."

Sam turned and faced the owner of the voice. "Ripe" was the word he would have used to describe her. She was in her very late fifties or very early sixties, but a heavy layer of makeup, just short of too much, kept Sam from making a clear distinction. She wore her long blond hair, obviously dyed, loose around her shoulders as a younger woman would have, and one look at her long eyelashes confirmed that she didn't believe in mascara—those puppies were false.

A simple black bustier that accentuated her more-than-modest proportions and a hip-hugging skirt that ended midcalf completed the package. Sam was sure she had left her feather boa draped over the back of a chair somewhere. The woman wasn't vainly clinging to her youth, he decided, so much as she was paying homage to a time when her features were more delicate.

"The groom can kiss my ass, and so can you," Tawny replied quietly. She stalked out the door and slammed it behind her. The older woman rounded on Sam.

"What did you say to her?"

"Nothing! I apologized."

"Okay." She gestured to the towel wrapped around Sam's head. "Finish your hair and be in the bar in five minutes."

Al had been watching the exchange wide-eyed, mumbling to himself, "Kathy Scherber-Danson . . . Kat Scherber . . . Kit Kat. . . ." Realization struck him as the woman sauntered away. "Kitty Kat! Sam, that's Kitty Kat!"

"Huh?"

"Kitty Kat! You know, the actress?" Al's expectant look turned into a frown when he saw that Sam clearly had no idea what he was talking about. "You're kidding me! Don't tell me you never saw *Kamikaze Vampire Sorceresses*." Sam looked blank. "How about *Die, Virgins, Die*?"

"Are those movies?"

Al raised his hands in frustration. "Of *course* they're movies! Great ones! From the golden age of sexploitation films."

37

Al's face took on a nostalgic cast. "I guess it *is* a little before your time, but in the late 1950s and early 1960s, Kat's movies helped transform American cinema."

"*Die, Virgins, Die*?"

"Don't scoff!" Al said seriously. "Films like that charted a lot of forbidden territory back then. Opened doors. They also almost single-handedly made the drive-in possible."

Sam shook his head. "Cheesy B-movies did all that?"

"Technically, they were B-movies. I don't think Kat's ever went above C-minus, though. And Dicky Danson. Who'd have thought a tomato like her would end up with a bit player like Dicky Danson?"

"What vital role did he play in our cinematic history?"

"Not much of one. He was always the geek who got killed in the first half-hour, after whining lines like 'But you *can't* go off and join the Chopper Vixens, we have a math test on Monday,' or '*Golly*, gang, maybe we should wait for the sheriff to get here before we check out that meteor crash.'"

Sam laughed. "So you're saying he was the voice of reason."

"The last thing you need when things are getting hot and heavy with your date in the backseat at the drive-in is a movie expounding reason," Al shot back. "I think Dicky's best work was that part in *Carnival Hell*, when the freakazoid gargoyle set his face on fire. He was very convincing as he ran down the street screaming."

Sam couldn't suppress a chuckle at that one. He was amazed at the endless store of inane knowledge his friend possessed. "So why am I here? To stop Richard from bursting into flame again?"

"Let's hope not," said the Observer, returning to his study of the hand link. "Ziggy still—"

The sound of shattering glass cut Al off. The two men were quickly on the move, Al relocating and Sam running to the bar.

He was relieved to see that no one was hurt. It could easily have been otherwise, though, judging by the size of the rock that was surrounded by the shards of what had once been the top pane of glass in the club's front door.

Kat brushed past him in a fury, almost knocking him over.

"What the hell . . ." was all he heard before she was out of earshot. He trailed her into the parking lot, but stopped short when he was met by an angry mob of picketers parading back and forth, signs raised.

Sam joined Al, who stood a bit to the side, puffing his cigar pensively, hand link all but forgotten. "Ziggy might be on the fritz," he said, "but I think we just found out why you're here."

CHAPTER FOUR

"What do we want?"
"Decency!"
"When do we want it?"
"Now!"

A crowd was picketing in the parking lot, chanting their mantra over and over. They looked to Sam like a pretty good cross section of people: bearded men in overalls and baseball caps (almost the required uniform for farmers) marched next to men and women in expensive casual clothes that said their work with the soil didn't go much farther than paying someone to keep their lawns tidy.

Most carried signs bearing slogans like *Stop the Slop!* and *No More Mud!* Others read *I Believe in the Rural Purity League* or, more simply, *RPL*.

Sam's gaze fell on a thin man with white hair and glasses who walked back and forth in front of the crowd. If his strut didn't prove he was the head of the group, his bullhorn clearly did. He wore an expensive gray suit with a blue tie that brought to Sam's mind flashes of Jeb Olsen, the longtime mayor of Elk Ridge, Indiana. Jeb had carried himself in a friendly but puffed-up sort of way, and he knew how to play the townspeople like a fiddle when he needed to.

This man was a bit slimmer, and clearly not as open, but he looked as if he could have given old Jeb a run for his money. People from the media were already on the scene, and the man strutted in front of the cameras like a seasoned pro.

"Who normally cavorts in mud, people?" the man shouted into his bullhorn.

"Pigs!" the crowd replied.

"Can we blame the pig for staying in its natural habitat?"

"No!"

"'No' is right," the man continued into the bullhorn. "But here we have humans doing the same! Why, this mud hole is not only an insult to our town, it's even an insult to pigs!"

The crowd roared in approval, marching double-time.

"Let's stop the slop, for men, women, and pigs everywhere!" the man continued.

"Stop the slop! Stop the slop!" The crowd parroted, caught up in the moment.

Kat stalked forward. "I should have known it was you, Barrenger," she practically growled. She planted herself in the man's path, hands on hips. Barrenger didn't stop until the two were almost nose to nose, determined look matching determined look.

"Al," Sam said to the hologram.

"Two steps ahead of you," the Observer replied. "That's right, Gooshie, Barrenger," he said to no one Sam could see. "See what you can come up with on the Rural Purity League, too."

"Just what in the hell do you think you're doing?" Kat asked.

"We're protesting the degradation of our town by your filthy establishment." The man stepped back and raised his bullhorn to the crowd. "Will we stand for it anymore, people?"

"No!" the crowd roared back. Sam's view of the confrontation was becoming obscured by the group of reporters gathering around them, microphones and cameras vying for sound bites and video.

"*First* of all, I doubt that two in twenty of this ignorant mob even know what the word 'degradation' means," Kat said icily. "And *second* of all"—she held up the rock that had shattered the glass in the front door—"how'd you plan on shutting me down? By killing one of my patrons?"

Barrenger didn't miss a beat. "That rock is just a symbol of the rage our townspeople feel toward your tawdry establishment." He raised the bullhorn once again. "Right, people?"

"Right!" responded many voices.

Kat grabbed the bell of the horn and pulled it out of Barrenger's hands. Raising it to her mouth, she said, "Well, you can all tell it to the judge. The police are on their way."

The effect was almost instantaneous; the crowd thinned to almost nothing as people disappeared into the night.

"That's just what I thought." Kat lowered the horn and gave it back to Barrenger with a satisfied smile. "I guess you'll be sticking around long enough to get thrown in the clink. Makes for good press, right? Dedicated to the cause and all?"

Barrenger turned toward what was left of the group, trying to rally the ragged strands that remained. "Stand firm, people! Stand up for your beliefs! Show them that our crusade goes deeper than grabbing headlines!" He started to march back and forth again, retrieving one of the signs that a less dedicated rabble-rouser had dropped during retreat. "She's made a life starring in sleazy freak shows, people. Do we want that type here, corrupting our quiet ways, twisting our youth?"

"That 'type,' Barrenger? That '*type*?' " Kat's eyes narrowed as she grinned slightly, her features becoming as feline as her name. "What *type* is that? The *wrong* type? Why, I do believe you've forgotten your white hood."

"Don't try and confuse the issue, madam. Don't try and confuse the people about the truth." Barrenger turned his appeal on reporters. "You see, don't you, folks? In the face of decency she has only one tactic to rely on." He turned back to Kat. "Well, try and make me look bad if you can.

Paint the worst picture that you know how. I have faith in our townspeople. Faith that they can recognize the truth when they see it. Why, I was born and bred here in Wilson, and I consider it my obligation and privilege to serve the fine people of this community."

Sam stopped listening. He had heard enough to know where this was going. Never mind giving old Jeb a run for his money; compared with old Jeb, Barrenger was a guru. Sam leaned toward Al, trying not to be too obvious. Not that it made much difference; all eyes, natural and electric, were focused on Barrenger as he paced in front of the gathering.

"This guy is *good*," he said. "I hope I don't have to confront him publicly to get through this Leap. It sounds like he could talk circles around an auctioneer. What's Ziggy say?... Al?" Sam shifted his attention from Barrenger to his friend.

The Observer was batting the hand link. He clearly hadn't heard a word Sam had said. "Dammit, Gooshie! I don't care what she says the problem is! You have the guy's *name*, for Pete's sake."

A flurry of chirping erupted from the hand link, and Al smiled. "That did it." He looked at Sam. "She never could take criticism. Our goose-stepping friend is Rex Barrenger. He's Wilson's Mr. All-America. He serves on the town council, and recently he's made it his pet project to shut down Kat's bar."

"Why? I can't say the entertainment's my cup of tea, exactly, but it seems pretty harmless."

"Well, the reason is less than selfless. He's using it as a trump card in his mayoral campaign. As if he *needs* it; he's running unopposed. And Ziggy says this isn't the first incident where Barrenger's acted as ringleader."

Al began punching the glowing buttons furiously. The information now seemed to be coming faster than he could read it. "Slow it down, dammit!" The Observer grimaced and waved his cigar in an arc around his head. "Quit trying to *overcompensate*!" He stopped flailing and made a satisfied grunt. "That's better. *Mudon*, Ziggy! Forget about your ego for a few seconds and just do your job. And I don't want to hear any more about Ann-Marie!" Al continued poking the hand link, sorting the data the computer fed him, but from the way he cocked his head, he was on the receiving end of a comment Sam couldn't hear. "Oh, yeah? Well you're more trouble than any of my wives *ever* were."

Sam was at the end of his patience. "Are you two done?"

Al listened again, smirking. "Her majesty wants me to tell you she apologizes." The hand link chirped in affirmation. "Brownnoser," the Observer mumbled.

"Al!"

"Calm down," Al said, trying to appease his friend. "Barrenger is backed by this Rural Purity League. The local newspaper describes it as kind of a consortium—Barrenger supporters who have the same, shall we say, *sensibilities*. But it's not clear who needs who more. Without League support, Barrenger probably wouldn't have the funds he needs for his campaign. But without Barrenger as its mouthpiece, there wouldn't be enough interest in the League for it to exist."

"How fortunate for them to have found each other," Sam said sarcastically.

"Oh, it's a cozy relationship for sure," Al continued. "With the League to legitimize his authority, coupled with his seat on the town council, Barrenger's been citing the club as the root of almost all of the town's woes. According to him, it's the cause of everything from increased drug use to the decline of morals and the erosion of family values."

"I know something about small town life, Al," Sam said. "I think it would be a little hard for him to make generalizations like that stick."

"On the contrary," Al continued, "the townspeople are eating it up. There's a presidential election this year, and the war on drugs is one of the hottest issues. So is family values. People are convinced that the country is in a bad way.

"Drug use has been on the rise everywhere, including here in Wilson. A local high school student died of an overdose this past summer, according to the paper." Al scowled at that bit of information.

"Anyway," he continued, "it's created the perfect little petri dish for Barrenger to grow his idealistic mold in. The club is cheap, the club is tawdry, the club doesn't fit in with the conservative 1950s ideology Barrenger is selling as a cure-all."

"How can people be so blind?" Sam said.

"Well, the town has gotten a bit of a swelled head recently," the Observer said. "Some national magazine rated it as one of the top five places to live in the entire U.S.— good school system with advanced academic programs in place, excellent medical facilities; even the local farmers have had slightly better luck than others around the country."

"How could something that positive lead to this nonsense?"

"Well, they didn't count on the bad stuff that's bound to come along with the good. Drug use is up because the population is up. Fact is, that kid probably would have died whether or not that article was printed.

"People from the cities are coming here in droves to raise their kids in a safe, small town environment," Al said, puffing his cigar. "But each new family that comes in changes that environment a little more. I guess you *can* call it the 'erosion of traditional values' if you like. The simple fact is that this town has been introduced to as many new ideas and different ways of doing things in the last few years as would normally happen in a generation."

"So the locals are a bit shell-shocked." Sam nodded. "But they must have been progressive thinkers in the first place to attract so many new residents, wouldn't they? This reaction seems a little extreme."

"I'm sure that at one point Dr. Frankenstein was also a progressive thinker with plans that looked great on paper. The locals are simply afraid of change. And as for Kat's bar being the focus of their anger. . . . Well, it's always easier to pin your problems on one demon. Especially one that you can see so clearly. Barrenger is relying on that mentality; come the beginning of next month, it lands him the mayor's seat." Al shook his head. "Small towns. . . ."

Sam shook his head, too. "*Not* small towns. *This* town." The cause of his worsening mood finally struck him. From almost the first day he had left Elk Ridge, he had found himself fighting the conventional wisdom that said people in small towns were ignorant, closed-minded hicks, even (and sometimes especially) with Al. While it *was* true that clubs like Kat's didn't ordinarily get much outward praise where he came from, they weren't harassed as long as they remained discreet. The people of Wilson were letting him down in a way, playing into the old stereotype. "I'm sure that the thirty or so people who were here tonight don't represent the majority," Sam continued, more defensively than he would have liked. "They were only the handful that Barrenger could rile up."

"Okay, okay," Al said, raising his hands in surrender. "I forgot who I was dealing with. I'm sorry for making generalizations. Geez." He lowered the hand link and squinted at the readout as more information came up. "But the fact remains that Barrenger gets elected and the club gets shut down. Oh, this is getting worse, Sam...."

"What?"

"Ziggy says that when the club closed down in the original history, most of the girls who work here wound up with less than kind fates."

"Less than kind?"

"Dreams shattered, opportunities gone, you name it." Al shook his head once more. "Each of the girls has her own story and her own reason for working here. Some are single mothers using the job to support their families, others are working their way through school and this is the only job they can find that pays enough."

Sam could tell his friend was hedging. Why? "What else is there, Al?"

"Well as clichéd as it sounds, the club is also a haven for a few of them. Steady work, a safe and stable environment"—he gestured at Kat, who was talking quietly with Richard by the ruined front door—"parental figures. Without the guidance of Kitty and Dicky, they're lost. AIDS, hooking, drugs...."

"Sarah is one of them, right?"

The Observer nodded.

"Damn! What can I do?"

"Ziggy says you need to stop Barrenger from shutting down the club."

"*That's* pretty obvious," Sam said, looking again at the man marching back and forth. "But how? He obviously won't change his position, not if he's using it to get elected and his continued funding depends on it. And I doubt I can get him to drop out of the race."

Sam began to pace, mirroring Barrenger. He didn't mind tough problems. Most times, in fact, he welcomed them. They gave Leaping more meaning. They helped him feel like he was *really* making a difference, not just acting as a time-drifting quick fixer.

But the most likely courses of action (however *unlikely* they seemed at times) usually presented themselves. Nothing was coming to mind now.

"I've got Ziggy working on different scenarios," Al said. "She should have some answers if...." He looked into space and raised his voice slightly, directing the last part of the comment elsewhere. "If she ever gets her act together!"

Here we go again! Sam laughed in frustration, and his pacing became more agitated. Maybe he took Ziggy for granted most times, but why was there always something wrong with her just when he needed her most? Of course, when it came down to it, asking that question was a more roundabout way of questioning his own supposed brilliance. He *did* design the parallel-hybrid computer, after all.

Was the fact that he had to rely on his gut instinct most times simply his comeuppance from God or Fate or Whatever—the boy wonder's slap on the wrist for deciding he could screw around with such a fundamental element as time?

Hubris theories aside, he was beginning to realize more and more that he was responsible for his own actions, and that his mental safety nets like Ziggy and Al and the Project were becoming less relevant as the years of Leaping went on.

His pacing took him past the corner of the building and his musings ended abruptly. There were two figures down the alley, talking in the pool of light that came through the club's open side door.

He didn't know who the man was, but the woman was definitely Tawny. And her mood didn't seem to have improved much. The two were arguing loudly enough that Sam could hear their voices, but he couldn't quite make out what they were saying.

"Al"—he gestured for the Observer to come look, and pointed down the alley. "What can you tell me about this?" Al punched the hand link and repositioned himself next to Sam. From the constant *blip-blip* the hand link was making and Al's continued silence, Sam knew the answer wasn't forthcoming.

"Listen," Sam heard Tawny say clearly, "this conversation is over. I don't know how you found me, but just leave me alone!" She began to stalk off, but before she had gone more than a couple of steps, the man grabbed her from behind.

"Oh, no," he said just as loudly, "you don't get to walk away from it all again. Not this time."

"Let go!" Tawny struggled in the man's grip. Sam didn't need Ziggy to tell him what to do.

He ran down the alley, grabbed the man by the shoulder, and yanked. The man spun around, letting Tawny go. Before he could do more than be surprised, Sam's foot connected with the side of his face, and he crumpled to the ground with a grunt.

"Oh hell!" Tawny cried. She dropped to her knees and rolled the man over, cradling his head in the crook of her arm. The look she shot Sam was ten times more deadly than the kick he had just given. "What the hell *is* it with you and this kung fu crap tonight?"

The man moaned, and his eyes began to focus. He looked from Tawny to Sam and back. "What the . . . ?"

"What do you mean?" Sam said, confused. "I was trying to help."

Tawny sighed. "I never *asked* for your help. But since you insist on butting in *anyway*"—she gave Sam a sardonic look that matched her voice—"I guess introductions are in order."

She lifted the man's head slightly and raised her voice. "Kyle"—his eyes struggled to meet hers—"the girl who just decked you is Candy, also known as Sarah." She wiped at the blood that was starting to trickle out of his nose. "*Ter-*

49

rific," she said under her breath. She turned her attention back to Sam. "Sarah, the man lying here bleeding all over me is Kyle."

"You know this creep?" Sam asked, starting to feel defensive again. Could he do nothing right with this woman?

"All my life," Tawny replied. "He's not just *any* creep. He's my brother."

CHAPTER FIVE

"Was your family happy durin' the holidays?"

Ann-Marie looked up from her turkey dinner and studied her friend. Evangelene never really came out and asked questions like that. Some things just weren't done, no matter how close a friendship got; questions that solicited personal information not given freely fell into that category.

Not that Ann-Marie hadn't asked a question or two early on. She'd gotten an answer the first time, but in a tone that almost made her forget what she'd asked.

The second time was the last. Remembering it still made Ann-Marie's guts clench, even though it had been . . . how long? *God, that was six years ago, and I still want to cry when I think of it.*

She and Evangelene had been playing cards in Ann-Marie's cell. It was only about a week after she had had her outburst (she couldn't bring herself to admit that it was really a tantrum), and the others on the cell block seemed about a step short of forming a lynch mob. Some of the offhand comments a few of them made while she was in earshot made her lie awake at night, thanking God she was safely tucked into her locked cage and wondering how she was going to spend the next ten years constantly watching her back.

Of course, the anger had faded with time. Only one inmate, that moron Rosa, still wouldn't let it go. Why, Ann-Marie didn't know. Maybe she really *did* have something going with Tibor. How Ann-Marie posed a threat to that, she couldn't figure out. But one look into Rosa's eyes confirmed that something had snapped up there, or was very close to it. Ann-Marie steered clear of her as much as she could.

But Evangelene hadn't been angry back then. On the contrary, she seemed almost desperate to form some sort of relationship. Rather than question it, Ann-Marie took advantage of the situation. The card games lasted only a few hands at first, but by the end of the week they went on for hours.

The first few times were more like lectures than social events. Evangelene told Ann-Marie who to approach for what, who to stay away from at all costs, and the rules and codes that passed for inmate etiquette.

Ann-Marie's question was harmless, really. More cliché than real dialogue. "So, what are you in for?"

Evangelene stopped in middraw and pinned Ann-Marie with a look that made her throat go dry. Anger and pain seemed to be fighting for control of her face.

Ann-Marie didn't know what to say. "I'm sorry."

Without a word, Evangelene placed the cards neatly on the bunk and walked out of the cell. Ann-Marie heard the door next to hers slam. It wouldn't have been too bad if that had been the end of it. But Evangelene hadn't joined the dinner line. And later that night, some weird twist of acoustics carried faint sobs into Ann-Marie's cell.

She'd never questioned Evangelene again. So she was surprised at her friend's query. "I don't know," she replied. "I guess we were happy. Christmas was always better than Thanksgiving. No football. Presents. I think I miss the Thanksgivings more, though." Ann-Marie looked down at her institutional holiday feast. The gravy had puddled into a crater in the mashed potatoes and was starting to congeal.

She mixed it up and tried not to think about it as it went into her mouth. "The meals, anyhow," she said after she swallowed.

"This dinner ain't that bad," Evangelene said. "Today's not supposed to be about food, anyway, it's about realizin' what we have and bein' thankful for it."

Ann-Marie almost choked on her peas. *Thankful!?* What did she have to be thankful for except a damp cell and seven years of wondering? Seven years of the mystery that was eating her up inside. Lurking. Forever living like sandpaper just under her skin, rubbing her raw from the inside out.

And here 1982 was sliding into 1983, the sameness of prison life turning it all into one mindless grind, and still no magic revelation. No justice. Not even a *chance* for justice.

Ann-Marie noticed the white-knuckled grip she had on her fork and realized she was scowling into her tray. *I will not let this get the best of me.* She took a deep breath. *I am in control.* She gently placed her fork down. *I am not an animal.* She grabbed her napkin and began folding, mindlessly producing a paper swan.

"So, was *your* family happy during the holidays?" To hell with etiquette. Evangelene had brought the whole thing up. It was fair game now. And she needed the distraction.

"Oh, yes," Evangelene replied eagerly. "We never really had much. Jim Crow wouldn't let us. But Daddy would always find a way to get all seven of us somethin' on Christmas mornin'. Then we'd go to services. Daddy'd say, 'I've rendered unto you what is yours, now it's time to render unto the Lord what's His.'"

"Sounds like we celebrated differently," Ann-Marie said. "We never went near church on Christmas, or any other time of year, for that matter. Christmas was about the tree and the eggnog and the stupid TV specials."

"My Thomas loved the tree"—Evangelene made the sign of the cross absently as she spoke—"and goin' to town to look at the shop windows. We couldn't afford a television. Not back then. There wasn't nothin' on for us, anyway. Goin' to church was the way for our community to share the Spirit."

Ann-Marie couldn't imagine a Christmas about brotherhood. It worked in theory (and in all those TV specials), but everybody knew the reason for Christmas was presents. End of story. As she got older, the presents got better (mainly because she bought her own gifts) and didn't always fit under the tree. Screw that sentimental crap.

But she offered a smile to Evangeline and pretended she understood. Why did people have this obsession with believing in something they couldn't see? There was only you. To put faith in anything else was foolish. It was the only weakness Ann-Marie had ever found in Evangelene.

"Well"—Ann-Marie raised her cup—"I *am* thankful that we're friends. I know I would have a much harder time in this place without you." *Not that it makes being cheated out of my life any easier*, she added as a mental afterthought.

She drank the last of her coffee—the dregs were bitter, which seemed fitting—then crumpled her swan into what was left on her tray. Happy Thanksgiving was over. Merry Christmas was next, followed by Happy Friggin' New Year. Only three more to go after this one.

The bell rang soon after, and Ann-Marie and Evangelene picked up their trays and joined the line back to the block. Fresh thoughts of the old mystery had Ann-Marie so wrapped in thought that she didn't notice a chair sticking out at the end of the table. Before she realized what was happening, she had tripped over it and was falling, instinctively grabbing at the women in front of her.

When she finally regained her balance, she looked at the shirt she was clutching, then at the owner. "Oh, shit...."

"Get off me, bitch!" Rosa twisted her torso, her eyes taking on a slightly deranged cast as she struggled to get free from Ann-Marie. This threw Ann-Marie off balance again, causing her to hold on tighter.

"Wait. Wait . . ." was all Ann-Marie got out before Rosa swung the edge of her metal tray into Ann-Marie's temple. Falling down wasn't nearly as scary after that.

She could sense a brightness behind her eyelids before she was able to open them. The air held a slight chill, raising goose bumps. Her mouth felt like it was coated with dust, and her head throbbed.

Where am I? Ann-Marie sat up with a start, hugging herself and attempting to rub some warmth into her arms. Flecks danced in her vision but soon cleared to reveal . . . nothing. That she was in a room of some sort was evident only because there was a floor at her feet. Aside from that, it was unlike any room she had ever seen. The walls seemed composed of a dull blue nothingness: no corners, no evident ceiling.

It may have been some god's palace in the sky, yet there was nothing regal about it. Aside from the cot she sat on, the only other bit of furniture she could see was a . . . "console" was the only word that came to mind to describe the narrow white table set into the slightly raised dias off to one side.

She spun around; the only other thing she could see was a silver-gray door at the top of a shallow ramp on the wall opposite the console. It helped to dispel the illusion that she was in midair, but the place was still eerie.

Her heart was beating faster, keeping time with her breath. She was still cold, and she felt naked. A look down showed that she was in a skintight suit that left only her feet and hands bare. *Where am I? What's happening?*

She approached the door, the only way out. *Out to where?* A dull whine filled the room, and the door began to rise. She ran back toward the bunk, her panic increasing. She was crouching on the bed, arms in a defensive pose, by the time the door was completely open.

A woman in a lab coat walked down the ramp, trying to read a clipboard and smile reassuringly at Ann-Marie at the same time.

"Who are you? Where am I?"

The woman continued approaching. She appeared bird-like to Ann-Marie, not overly tall, but slender enough to give that impression. "I'm sorry if your head aches a bit, but we had to sedate you earlier." Her manner was professional; friendly but detached. Large eyes stared at Ann-Marie from her dark face.

Earlier. . . . The last thing Ann-Marie could remember was seeing two people, dressed like this one, and someone screaming. "What's going on here?"

"That may be a bit difficult to explain," the woman said, still approaching, "but you're safe. No one here will hurt you."

"Damn right!" Ann-Marie sprung from the bunk, launching herself at the woman . . . *to do what?* She had a vague sense that she should be able to flip this woman or kick her or something, but she couldn't remember how. She stopped short, just in front of the woman.

The woman took her by the shoulders and gently pushed her back on the bunk. "I promise, no one will hurt you." She tried to lay Ann-Marie down once again, hands gentle but firm on her shoulders. "I'm Dr. Beeks. I'm here to—"

"I don't care what you're here for!" Ann-Marie began to struggle harder. "Where am I?"

"Please calm down," Beeks said, matching Ann-Marie's increasing strength. She lifted her head. "Ziggy, get me some help!"

Ann-Marie continued to fight as the other woman's eyes filled her vision. She squeezed her eyes shut and pushed and. . . .

Reality shifted.

The floor was hard and cold against her back. Ann-Marie opened her eyes and Evangelene stared back at her, concern filling her features.

"She's crazy! The bitch is crazy!"

Ann-Marie looked past Evangelene and saw Rosa, held by Hernandez, staring down at her with fevered eyes. "Look at her, rolling around down there!" Rosa continued. "I had to hit her. She attacked me."

"She tripped," Evangelene said levelly. "It was an accident."

"But Evangelene, Rosa says. . . ."

"Mr. Hernandez"—Evangelene raised her voice slightly—"have you ever known me to lie?"

"No," the guard admitted.

"Then why are you callin' me a liar?" She turned back to Ann-Marie and rubbed the side of her head. Stars exploded behind Ann-Marie's eyes and blackness coiled on the edges of her vision. "The matter was all very simple. This woman tripped, and that . . . *woman* attacked her for it. I'm *sure* others will agree."

Some of the other inmates nodded. Most just studied the floor. Hernandez looked from Rosa to Ann-Marie and back, then shook his head. "Accidents happen, I suppose," he said finally. "Fall in and get back to the block." He released Rosa and swept his gaze across the inmates clustered around him. "*Now*, ladies."

The others finally began to move. Evangelene helped Ann-Marie up and both joined the line. "That Rosa," Evangelene chuckled. "The way she scowls . . . like she was 'spectin' a chicken dinner and discovered there was nothing left but jowls and beaks. Well, as my daddy used to say, 'Chicken one day, feathers the next.' "

Beaks? The word swam through Ann-Marie's head as she rubbed absently at the potatoes and gravy smeared across the front of her clothes. But the pain clouded everything. She struggled to remember, but the only image breaking through the haze was an open blue sky. She knew it meant more, but comprehension danced just beyond her reach.

She was back on the block before she realized it, and in a worse mood than before. When Evangelene asked to join her in her cell, she grumbled assent.

"That cut needs tendin' to," Evangelene said, making her way to the sink. "Cool cloth will have to do."

Ann-Marie sat on her bunk and retrieved the cards from the bed table. She was shuffling absently when Evangelene applied the damp cloth, dabbing at the bloody lump. "This will set things right in no time."

"*Nothing* can set things right!" Ann-Marie threw the cards down and stood up, pushing Evangelene's hand away. "Don't you understand that?"

"Come, now. It's only a small bump."

"I don't mean this, dammit!" Ann-Marie gestured at the lump and began to pace. "I mean *this*!" She waved her arm in an all-inclusive gesture. "This hole. This *life*. I don't know how it happened, and every time I think I might be close to figuring it out, I can't remember."

Evangelene looked at her quizzically. "I don't see how...."

"Don't try! You wouldn't understand," Ann-Marie sat next to her friend and took deep breaths. "The injustice of it all . . . I just don't belong here."

Evangelene stood suddenly, facing Ann-Marie with a look that withered her. She thought she had seen Evangelene mad before, but compared to this, the woman had only been mildly annoyed.

"You have the *nerve* to talk of injustice?" Evangelene bit off each word as if she were struggling to contain a whirlwind within her. "From the first day, I have listened to you whine and prattle about how you don't know how you wound up here. The injustice of it all. Well, you don't know what injustice *is*. You can't even begin to comprehend it."

"But. . . ."

"But nothin'! It all seems very simple to me. You robbed people. You got caught. You went to jail. It's black and white. Black and white. . . ." She laughed bitterly, a sound of rage and pain; it disturbed Ann-Marie the most of anything so far. "It's black and white put me in here, too," she continued. "Black and white that killed my Thomas." The laugh turned into a sob. Evangelene sat down next to Ann-Marie.

"They came in the night, the cowards. 'Bout twenty of 'em all hollerin' and laughin' and burnin' a cross. *A cross*! Our house was next. I heard my baby screamin', but I couldn't get to him in time. Barely got out myself."

Evangelene stood once again, as if moving toward something she saw in the distance. "I showed 'em, though." The smile was back. "They had the *gall* to march in the Independence Day parade. Said it was their right as free Americans to express their views. First Amendment. So I figured it applied to me, too. I got three of the bastards before they tackled me and took the rifle away."

Ann-Marie sat in stunned silence.

"Didn't give me death, though," Evangelene continued. "Even changed the venue of the hearin'. It hit too close to home. Too close to their dirty little secrets. Judge said, considerin' the mitigatin' circumstances, justice would be served with a life sentence." Evangelene laughed again. "*Justice would be served.*" She shook her head and was silent.

Ann-Marie stood and put her arms around her friend's shoulders. "You poor thing!"

Evangelene shrugged her off and put a few steps between them. "Oh, no, I'm no victim. I will not sully Thomas's memory by calling myself such." She turned and looked Ann-Marie in the eyes. "I did what I *could* do. The only thing I saw to do. That doesn't mean it was right. Nothin' can make it that. And even knowin' it, I'd do it again. So everyone got what they deserved. Eventually, *everyone* does."

Ann-Marie sensed that the subject had turned back to her. It was her turn to start pacing. "But how do you cope, when it's always right there, shadowing everything you do?"

"You find things to fill the time, to fill the hole."

Ann-Marie swept the origami figures off the table and kicked them across the floor. "But folding paper just isn't cutting it anymore!" she cried. "It's not enough."

"I know you're more intelligent than that," Evangelene shot back. "I saw it from the first. Why do you think I bothered talkin' with you? I got along in here for years just fine without you. But I saw somethin' in you. Call it a spark or whatever. Someone who could maybe hold a decent conversation. Even if we wound up hatin' each other, I couldn't sit back and watch this place drain the life out of another person. *Create* another victim." She sighed. "Maybe I'm not as good a judge of character as I thought, though. From what's comin' out of that mouth of yours, you were a victim long before you got here."

Evangelene knelt and picked up the folded figures at her feet. "It was never about paper or foldin'. It was about showin' yourself that you're better than your circumstances. That you control your inner outlook, and to hell with outside events. You knew that in the beginnin'. But you lost it somewhere along the way. You let the years and the supposed *injustice* leach the meanin' out of it."

Ann-Marie leaned on the sink, frustration rising. "I want to begin again. I want to keep fighting."

"Then do it," Evangelene said. "You can't wait for me

to hold your hand and lead you through it. You have to find your own way."

"How?"

"I don't know. I wish you'd let the Lord in. It would help so much"—Evangelene raised her hands in surrender to Ann-Marie's stony glare—"but that, most of all, is somethin' you need to figure out for yourself. Just find somethin' that you can focus on. Somethin' that you can devote yourself to. What it is doesn't really matter, as long as it helps you get in touch with who you are and who you want to be."

Ann-Marie nodded. Evangelene made it all sound so simple. "I have to try," she said. But even as she uttered those words, she knew they contained more than a hint of lip service. Try as she might, she doubted she would ever be able to put the mystery completely to rest.

CHAPTER SIX

Sam hooked his toe behind the alley door and drew it open. He needed both hands to hold Kyle up. He must have given the guy more of a kick than he thought.

Red and blue lights penetrated the darkness of the alley, throwing the club's brick wall into staccato relief.

"What's that?" Tawny looked toward the mouth of the narrow passageway. "Cops? What are they doing here?"

"There was a little excitement going on before," Sam said. The door slipped away from his foot and slammed shut. "Damn! Can I have a little help here?"

Tawny pulled the door open. "This night is just full of surprises, isn't it?"

"Well," Sam began, readjusting Kyle's weight and getting his shoulder behind the door, "Barrenger is at it again and—"

"I don't really care." Tawny turned her attention to her brother. "Kyle? You okay?" The man's head lolled a bit to one side. "Good. Now stay out of my life! And you, too," she added to Sam. She began to walk away.

"Wait!" Sam called. "Aren't you going to give me some help?"

"No. Why don't you two bother *each other* for a while?"

Sam sighed. "It's just a little back this way," he said, hoisting the man through the short hallway that led to the dressing area.

"Too bad the girls are gone," Al said. "I bet a look at all the goodies would snap him out of it."

"Can't *you* be a bit more help?"

Al gave his cigar a pensive puff. "Not really."

Sam shook his head and lowered Kyle to the floor so he could open the door to the lounge. He then dragged Kyle into the room and laid the unconscious man on the nearest couch.

"This is Kyle Singer," Al said as Sam put the man's legs up. "Geez, look at him. He's as Irish as Paddy's pig."

"What can you tell me that I can't find out by looking myself?"

Al stopped his study of the man and shifted his attention back to the hand link. "He's an investigative repo..."—he whacked the hand link—"repo..."—*whack*—"reporter." The Observer nodded. "Bachelor of Journalism from the University of Missouri in '86. Works for a television station in Paducah, Kentucky, now."

"Mizzou?" Sam paused. "That's where Katie went to school." He couldn't remember what his sister got her degree in, but he was pretty sure it wasn't journalism. "That's where she met Jim. I *think*."

"You're right," Al said, continuing his quest for information on Kyle. Sam wished the Observer would elaborate more on Katie. He remembered that Jim Bonnick was her husband and that he was in one of the branches of the military and . . . nothing else.

"From what Ziggy can find," Al continued, "Kyle's doing fairly well making a name for himself. He heads an investigative team that's turned out some pretty solid work."

Sam got a wet towel from the bathroom and laid it on Kyle's face. The nosebleed had stopped, but a lump was forming on his cheekbone. "What's he doing here?"

"Seems pretty obvious. He's looking for his sister."

"Think so?"

"Think what?" Kyle rolled his head slightly and struggled to sit up. Sam surrendered the towel and took a few steps back.

"I said I think you're coming around." Sam tried to smile nonthreateningly.

Kyle swung his feet to the floor and shook his head. "Mistake," he groaned, and pressed the cool towel to his face again.

"Sorry about that," Sam said. "I didn't know who you were . . . and a girl can never be too careful."

"Neither can a man, apparently," Kyle said, trying to focus on Sam and wipe the tacky blood off his upper lip. "Not around you, anyway. And you are . . . ?"

"Sarah," Sam replied. "Let me see if I can find some ice for that."

"No, no. I'll live." Kyle stood to his full lanky height and took a step toward Sam, extending his hand. "Kyle Singer."

"I know." Sam took the man's hand. "Tawny told me."

"Tawny?"

"Your sister."

"Oh," Kyle shook his head and reeled back a step. "You mean Theresa. *Tawny!*" He uttered the word like a curse. "I can hear Mom spinning in her grave now. Well, Tawny or Theresa, thanks for sticking up for her."

"From what I've seen, it looks like she can stick up for herself," Sam said.

"Well, you'd probably know better than I would. We haven't seen each other in some time." Kyle sat down, gingerly testing the dark blue knot that was rising under his left eye. "Hell," he chuckled, "*there's* irony for you. I've finally found her, and with the way this is swelling, I *still* won't be able to see her."

"Bingo!" quipped Al.

"You've been looking for her?"

"Almost two years now," Kyle said. "I guess it might have been a little quicker if I knew to look for *Tawny*."

"Why haven't you seen her? What happened?"

"That's what I'd like to know. I came home from work

one night and she was gone. Left town. Left Dad. Didn't leave a note."

"Ziggy says 'Dad' is Robert Singer," Al chimed in. "According to hospital records, 'Mom' passed away in 1984 from a heart attack." He growled at the hand link. "Ziggy won't give me any more."

"She just disappeared," Kyle continued. "I've been looking for her ever since."

"Wow," Sam said. "Why would she do something like that?"

"I think that would be less of a mystery if you met my father. He's not exactly easy to live with. But that's still no excuse. She was supposed to take care of him until I got finished with school. That was the deal."

"Something might have happened," Sam tried. "With your father, maybe?"

Kyle shrugged. "I don't know, and at this point I don't really care. She obviously didn't."

"Well, now that you've found her, what are you going to do?"

"Bring her home to Kentucky, of course," Kyle said. "I refuse to let her do"—he furrowed his brow and looked around the small room—"... *this*. No offense," he said to Sam as an afterthought.

"None taken," Sam said in a tone that betrayed the lie. He was willing to give Kyle the benefit of the doubt after hearing his side of the story, but he couldn't condone a judgmental attitude (no matter that he probably would have reacted the same way if he found his *own* sister working as a mud wrestler). "I get the feeling Tawny might not be so forgiving, though," Sam went on. "Your attitude will probably make things worse."

"You got that right, Sam." A hint of disgust tinged the Observer's voice. "Anyway, you *may* have that right. Ziggy says Tawny stays here for another week or two, then pulls another vanishing act. This time it's for good. We may not be able to bet on that, though, not with the case of schizophrenia our obnoxious bucket of microchips seems to have come down with."

Kyle barked a laugh. "Please! How can things get any

worse? I spend all my spare time worrying about and looking for my sister. I finally find her—rolling around topless in the mud, mind you—and she wants nothing to do with me. And then I get decked!"

"I know you're frustrated right now"—Sam tried to placate the man—"but don't you think it might be better to *talk* to your sister instead of making demands?"

"Look," Kyle shot back, "I don't see how it's any of your business. I mean, who the hell are you, anyway? What do you care?"

"I'm . . . a friend," was all Sam could think to say. Kat walked in and prevented him from saying more. And if the harried way she was adjusting her bustier was any indication of her mood, Sam didn't *want* to say anything more. He wanted to run.

"Wow!" Al exclaimed. "She looks just like she did in *Revenge of the She-Beast*, in that scene where she chews the deputy's nose off! Look out, Sam!"

Sam stepped backward as Kat turned his way. "Thought you'd hide in here, huh?" The question was aimed at Kyle. "Well, you have two choices. Get outside, where the sheriff is waiting for you, or I kick your ass out there for you."

Kyle backed up until the wall stopped him. "Huh?"

"Kat, wait." Sam stepped between the two. "He's not from Barrenger's group."

Kat peered at Kyle with eyes that held a laser-blue intensity. "Then who the hell is he?"

"Tawny's brother."

Kat shifted her gaze to Sam. Butterflies rose in his stomach, then withered. "Tawny's brother? What the hell is he doing here?" She cocked her mouth in a wicked grin. "Get kicks watching your sister?"

Kyle took a step forward. "Now just a minute. . . ."

"Not now, Kyle!" Sam raised his arms between the two. This had the potential to get very ugly. "Not unless you want to be lying on that couch again." He focused on Kat. "He's been looking for her," he said. "They haven't seen each other for over a year."

"Two years," Kyle corrected.

"How touching," Kat said sarcastically. "Look, I don't

care if you just found your long-lost teddy bear. My club is closed. Take it up with your sister at her place or come back during business hours.''

Kyle straightened, raising his chin a bit. "I'll be back." He looked questioningly at the exit. Sam nodded. Kyle's long legs carried him out of the room in the space of a breath.

"A pleasure to meet you," Kat yelled as the door closed. "Let's do it again real soon." She stalked to the couch and sat. "From the shape his face is in, I'd say he hasn't figured out when to keep his mouth shut." She lit a cigarette and made an attempt at calm, but with the smoke coming out of her nose, she reminded Sam of a cartoon bull with the matador in her sights. "I'm getting too old for this crap. What was he doing in here?"

The room seemed to shrink around Sam as Kat's ire centered on him. He sat next to her and related the events of the evening. Kat chuckled when she heard how Sam had kicked Kyle. "Don't feel bad, honey. He's probably had it coming for a long time. Trust me."

"I don't know," Sam said in Kyle's defense. "I'd say he's been through a lot."

"Even if that's true, it doesn't matter," Kat said. "If there's anything I've learned over the years, it's not to get involved in family matters. Especially in this business. If I had a dime for all the irate fathers I've faced, I'd have enough to buy Barrenger's silence. What none of them ever seems to figure out is that their precious little babies are here for their own reasons; reasons that have a lot to do with the fathers in the first place." She took another drag on her cigarette and shook her head. "I tried to help the first few times I came up against a situation like this. I really did. But it only ends up bringing more problems, especially when people think you're meddling in their personal lives."

Sam nodded. He knew the feeling all too well. It was difficult enough to help people face their problems, even when he had the advantage of being perceived as a family member or close friend. It was always hardest when he was almost a total stranger to those he was helping. Like on this Leap.

"So *let* them think Richard and I are evil Gypsies roaming

from town to town luring their innocents away. It makes it easier for everyone."

Sam wasn't above using a little guilt if it would gain him some help. And from what his gut was telling him, getting Kyle and Tawny reconciled was just as important as keeping the club open. "Easier for everyone except the girls caught in the middle," he said.

Kat looked sharply at him and was on her feet again. "How *dare* you? When have Richard and I ever been anything but supportive of you—or anyone else in this club, for that matter?"

Well, that one blew up in my face.

"You have to draw a line," Kat continued. "If you don't, you have nothing but trouble." She crushed her cigarette into an ashtray. "And I have plenty of that at the moment."

Sam decided to switch gears. Kyle and Tawny could wait for now. "Barrenger's got you worried, doesn't he?"

Kat nodded. "At first I just laughed it off. I mean, who ever takes fanatics like that seriously? We never used to, and they were *always* crawling around the theaters where our movies were shown. This time it's not so easy."

By the slant of the creases around her eyes and the wrinkles in her forehead, Sam knew Kat was more than just worried; she was downright scared. It made her look older. "Did you see that mob?" she continued. "Sheep and hypocrites. Half those men come in here at least once every two weeks. The others didn't have any problems with me until Barrenger told them they did. And that rock-throwing bullshit? Someone could have gotten hurt." The anger was creeping back. "It's not dying down. It's getting worse."

"There must be something we can do," Sam said, "some way we can turn it around. People are just reacting out of fear. We need to make them see it for themselves."

"Fat chance," Kat replied. "Everyone has his own reason for being riled up, and I certainly can't afford to send the whole town to counseling."

An idea began tickling at the back of Sam's mind. *Counseling . . . counseling . . .* Council!

"I know what to do!" he exclaimed. "If we can't beat Barrenger by making a stand here, then why don't we face him on his own territory? Why don't we take our case to the next town council meeting and let them decide?"

How could he have been so blind? Back in Elk Ridge, the town council held a lot of clout; it was where all the decisions were made. It had to be the same in this town, or Barrenger wouldn't be so effective.

"Let me get this straight, honey." Kat's voice was laden with condescension. "The wolf is knocking at the door and you want me to let him in? Did you take one hit too many in the ring tonight?"

Sam shook his head. "Don't you get it? Sooner or later the whole thing will come down to a council vote. Why not beat Barrenger to the punch? It might give us the edge we need."

Kat looked hesitant. "And if it doesn't work?"

"What do we have to lose? At least this way we'll be putting Barrenger on the defensive instead of the other way around."

"*We* don't have anything to lose. Richard and I, on the other hand, have sunk a good deal of our savings into this club."

"If you let this continue the way it is much longer," Sam said softly, "you're gonna lose it all. If you're gonna go down, you might as well go down fighting."

"Funny, that's the first thing I taught you when you stepped into the ring." There was a glint in Kat's eye that hadn't been there before. "You've made your point, but we need to get on the agenda for the next meeting."

"That should be simple enough. All we need to do is contact the councilman for this district. They can get the ball rolling, schedule a hearing. Even if that turns out to be a wash, there should be time allotted for open remarks at the end of the meeting. When you come down to it, they have no choice *but* to listen to you if you want them to. Even Barrenger."

Kat's smile turned predatory. "We'll show that bastard." She stalked to the door. "I'm going to call the girls right now. We need to plan a strategy. I don't think I'll have much of a problem convincing them to come in on a Sunday, not if it means a continued paycheck. Be here at eleven tomorrow morning. And keep the ideas coming." She winked at Sam and was gone.

Sam laughed aloud and looked around for Al to see how his idea had affected the odds. But Al was nowhere in sight. Only then did Sam realize that the Observer had been uncharacteristically silent during his entire conversation with Kat. Where had he gone? The Imaging Chamber door hadn't opened.

He caught the sound of a faint *poing*. He followed it into the outer hall that led to the club's side door. Al was there, and he looked like he was having a fistfight with the hand link. From the scowl on his face, the hand link was giving as good as it was getting.

"What's going on?"

Al looked up in surprise. "Sam! What's going on?"

"Are you in an Imaging Chamber or an echo chamber?" Sam asked in frustration. He thrust his thumb over his shoulder. "Didn't you hear what just went on in there? What does Ziggy say?"

Al glared at the hand link. "I can't get her to say anything. I don't know what's going on, but this thing is giving me nothing but gibberish." He shook the multicolored unit, producing some unusual squeals. "What did you come up with?"

"The town council meeting," Sam said. "What will happen if we bring the case before the council?"

"Not a bad idea," Al said. "Look, we obviously have a glitch in the system somewhere. I'm sure it's nothing. Let me take care of it, and then I'll bring you the odds."

"It sure doesn't *sound* like nothing," Sam said. The look on Al's face was starting to worry him.

Al hit the glowing buttons in a familiar pattern. "Don't worry. We'll take care of it. I'll come back to you as soon as I . . ." The words died as the glowing white door stopped rising at Al's knees. "For the love of. . . . Gooshie! Open the damn door!" Al cocked his head for a beat, his face getting redder. "That's it!" He got down on his belly and began to squirm into the light. "When I get outta here, I'm gonna tear that computer a new—"

The hologram vanished from Sam's vision. He could picture Al smashing the hand link against the floor in frustration and breaking the neural link.

He had a sick feeling in his stomach. He couldn't remember Ziggy ever having problems *this* serious. Could the cause really be as simple as Al suggested? He hoped it was.

What else *could* he do?

"I see you've lost your fear of live shots, Ms. Dotorovic."

Nancy looked up in surprise at Kyle's mocking tone, then smiled.

"What the hell happened to your face?"

Kyle laughed. Same old Nancy. "Nice to see you, too. I see you're still not one for platitudes. What's this big story they have you chasing down?"

"It's just a routine protest."

Kyle hadn't seen his friend in more than two years, but he had worked with her enough in school to know she was hedging. "What do you mean, 'a routine protest'? Do people in this town often go to jail that easily?"

He didn't know what was going on, but he smelled a story the second he walked out of the alley and saw the police cars pulling away. Reporting instincts took over. The old bat *had* accused him of hiding and had mentioned the police.

The parking lot was deserted except for a news crew wrapping up, satellite remote antenna retracting amidst its orange coils into the top of the van, waiting for the promise of another breaking story to arouse it again.

When he caught sight of the van's logo, he thanked God and the alumni society. The final bit of luck fell into place when he saw Nancy wrapping up her microphone cord.

Her uncharacteristic silence and the scowls he was getting from her partner told him it was a bigger story than he had thought. Nancy continued staring at him, eyes widening slightly. He nodded, signaling he got her message.

Not here.

Kyle focused on the man giving him dirty looks and smiled insincerely. "Kyle Singer." He extended his hand.

"This is Jerry Browne," Nancy said hurriedly. "He works with me at the station. Jerry, Kyle and I went to school together."

"Another member of the Mizzou Mafia." Jerry took Kyle's hand, returning the smile with interest.

Kyle already hated him. "I see they work you to the ground here, just like at my station. What's this, your third story today?"

"Fourth," Nancy said. "We were clocking out when the call came in. At least it was a live shot. No *crash* editing."

"I don't think you have to worry about that tonight." Kyle laughed. "No snow." She could only be referring to the time they were sent to cover a story in Jefferson City during a blizzard, the likes of which only Missouri could produce. On the way back to the station, they slid into a ditch.

Jerry looked on in confusion, mouth quirking in anticipation of a smile. Neither bothered to let him in on the joke. Kyle found the idiotic look becoming.

"Old times," Kyle said. "When do you think we can catch up on new ones?"

Nancy's eyes spoke volumes. "Monday night's good for me."

Kyle nodded. "Monday it is."

After a quick exchange of phone numbers, they bid one another a good night.

Kyle's mind raced as he watched the van drive away. He had stumbled onto something huge, at least for the people of this town, and the club was right in the middle of it.

Could he use it to persuade Theresa to come home? As much as he hated to admit it, the blonde was right. He needed a new angle. Now that he had a source, finding it would be a bit easier. Research and legwork would take care of the rest. They always did. He stopped thinking like an older brother and started thinking like a journalist.

Treat the whole thing like a sticky interview; circle around and around until you find an opening, then move in for the kill.

CHAPTER SEVEN

She didn't know what it was. Sometimes it was like that. The pencil just did what it wanted to, regardless of her original intent.

She had chosen drawing because it was an active exercise, one that kept her thinking, kept her growing. There was no chance it could degenerate into a mindless habit, as the origami had. And to her surprise, she had gotten pretty good at it. When she drew a tree, it looked like a tree. Her self-portraits had also been pretty good likenesses.

But this was one of the weird drawings. She supposed the image could pass for a table, but it was curved and thin and angled like something out of one of those dumb outer-space shows her brother used to watch. It stood on a low dais. Something in her mind told her the absence of anything else was fitting; the drawing was finished.

"I don't know how you escaped from Art-Deco Hell," she said to the image on her pad, "but you can't hide out here." She ripped the page off.

She had wanted to attempt a nature scene like the one on

the Japanese screen in her office. When she *had* an office. When she had a *life*.

Stop it! You draw to build yourself up, not drag yourself down! You are a creative, intelligent human being. What's done is done. Concentrate on the future.

But instead of throwing the drawing away and forgetting about it, she unclenched her fist and smoothed it out. She got up and carefully placed it under her mattress. It joined a growing collection of artistic enigmas.

A bed that *wasn't* the one in her cell. A door that *wasn't* her cell door. The woman.

She studied the woman a lot. She was black, like Evangelene, but that was where the similarity ended. If it *was* just a botched sketch of her friend, there should have been *some* resemblance. No, this was a totally different person; one she felt she should know. Only she didn't know from where.

As odd as the sketch of the woman was, however, it wasn't the oddest of the lot.

Ann-Marie forced herself to shuffle the papers around until she found the drawing she was looking for. Every time she saw it, her heart leaped to her throat. It terrified her.

It had started innocently enough. She had drawn herself again, a profile this time. But the work had expanded of its own volition. Before she knew it, her head was attached to a body clothed in a jumpsuit much tighter than any of her prison garments. Floating in the air in front of her was a mirror. But the image in it wasn't her. That was the terrifying part.

Her reflection was an ill-defined, shadowy figure. There was a hole where the face should have been. But it was *more* than a hole; her pencil lacked the ability to portray it accurately. The hole tore at her, pulled her in, trying to clutch her soul and consume her identity.

She dropped the picture with a gasp, sweat rolling down her sides. She moved the other papers so they would cover it and dropped the mattress over them all.

Still, she couldn't stop thinking about it. All of it was connected somehow—all of it circled back to the secrets that prowled in the dark corners of her mind and made her wake up almost every morning, gasping in fear and confusion.

Forget about it.
I can't!
Try harder.
It's not working! It always comes back!
Don't let it. Bury it.
I don't even know what I'm supposed to bury!

Tears of rage and fear mingled with the sweat on Ann-Marie's cheeks as the mental argument continued.

It's a weakness.
Stop it!
Weakness is for fools.
Shut up!
Weak-minded fool.
ShutupShutupShutupShutupShutup

"Shut up!" The words tore from her throat and she choked on a sob. She ran to the sink in a fit of coughing and doused herself repeatedly. The cold water helped return calm and banish the voices, but she fell to her knees, sniffling and hiccuping through chattering teeth.

I am not *weak-minded. I am* not *imagining all of this.*

She grabbed her pencil and scored a blank sheet of paper with thick lines, marking the repetition of those words in her mind.

She coaxed her thoughts and hand to rest. The pencil's tip had long since shattered, and tiny slivers of lead and wood decorated the page, which was now torn through in places.

The adrenaline departed, leaving exhaustion in its wake. She threw the pencil at the wall, swept the pad to the floor, and curled up on her bunk.

Sleep took Ann-Marie, and the monsters lurking underneath the mattress soon did the same.

She fought at the straps biting into her wrists and ankles, arching her back in a vain attempt to get free. Her hair was plastered to her forehead and sweat stung her eyes. She collapsed, panting.

"Where are you, you bitch?" she yelled. "Come out!"

A low whine filled the room, and Ann-Marie craned her neck in the direction of the rising door. The doctor was back, but this time she wasn't alone. A large man in a military uniform accompanied her.

"Close enough, soldier," Beeks said, gesturing to the wall with her chin. "Stand post there. Move only on my word." The guard nodded and stood at ease beside the door. Ann-Marie thought he looked even more menacing in his relaxed stance.

"Ms. Renerie"—Beeks turned to Ann-Marie, smile fixed in place—"I really hate to see you like that. And I don't want to have to medicate you again. Will you promise to stay calm if I loosen your restraints?"

"How did you know my name?"

"I'll answer what questions I can if you promise to give me a bit of cooperation. Agreed?"

"Yes." Ann-Marie bit off the word. As a rule she permitted others to see weakness only when it was calculated to get her what she wanted. And while surrendering to this woman *would* buy Ann-Marie freedom, it wasn't on her own terms. It irked her.

"I don't know if I can take your word on that," Beeks replied, "but as you can see, I don't have to." She glanced meaningfully at the guard. "I *do* know you're terrified, but I *promise* no harm will come to you, for whatever it's worth."

Terrified?! We'll see who's terrified soon enough, bitch!

Ann-Marie smiled weakly. "It's worth a lot. I just want to know what's going on."

Beeks hesitated for an instant, then loosened the restraints. Ann-Marie sat up slowly, rubbing circulation back into her wrists as Beeks set her feet free. She stamped her heels alternately on the floor, banishing the pins and needles.

"Are you hungry? I can have food brought in."

Would she send the guard? "That would be nice."

Beeks nodded. "Ziggy? Have someone bring...." she looked askance at Ann-Marie.

What the hell? "Foie gras and squab?"

"Some paté and something to drink for our guest. Sorry, the duck's off today."

The guard remained where he was. "Who are you talking to?"

Beeks smiled. "We have a rather sophisticated intercom system here. I guess you could call Ziggy our dispatcher."

"Dispatcher *indeed*," said a voice from out of nowhere. Ann-Marie looked around, but couldn't locate its source. It seemed to come from everywhere. "I am capable of handling far more advanced tasks—"

"That's enough, Ziggy!" Beeks said curtly. "You know the rules. This conversation is over."

"Very well." Silence followed.

"She always has to get in the last word," Beeks said under her breath. "Never mind studying Visitors. I could write another thesis on her alone!"

"Is that what I am? A Visitor?"

"Yes," Beeks said. "That's how we refer to our guests."

"How did I get here?"

"You have become part of a... sociological experiment."

"What?"

"Your... persona... has been switched with that of a scientist, a doctor. He is now acting in your stead to observe human behavior."

"*What?*"

"To everyone else, this scientist is you. He looks like you. He talks like you. And for the experiment to be a success, he has to *act* like you." Beeks patted Ann-Marie's knee and her tone became soothing. "That's why I need to know more about you and what you do. As far as we can ascertain, you are in the import-export business. But we know there's more. And you need to tell us what it is if you want things to return to normal."

Ann-Marie laughed. The explanation was so ludicrous that it caused something to click in her mind. "What are you trying to pull, lady?" She became furious at herself for not thinking of it sooner.

They were deliberately trying to throw her off balance. To hell with the innocent act. She would have to move soon, guard or no guard. "Do you think I'm stupid? You're with the FBI. Colton gave you my name to save his own ass. You concocted this whole setup so I would confide in you. Well, you obviously have nothing solid on me or you wouldn't be going through all this."

Ann-Marie stood and made a gesture that encompassed the room, laughing again. "Look at this! And that crap about a sociology experiment and switching personas. It sounds like a sci-fi story my idiot brother would make up, only he would probably find a way to throw time travel into it somehow."

Beeks gasped at the last comment. That was all the confirmation Ann-Marie needed. "What's the matter? Cover blown? Now charge me or release me, but let's drop the act."

"I don't know who Colton is," Beeks said, regaining her composure, "but you're wrong. There *is* an experiment, and if you ever want to go home again, you'd better start cooperating."

"Shove it up your ass. I want to talk to a lawyer."

Beeks faced Ann-Marie, furious. "I will thank you not to use that kind of language. And if it's proof you need, I can provide it." The doctor walked to the console on the opposite side of the room. "I hesitated to do this before because you were so unstable, but there's no other way. Come here."

I'll show you unstable! Ann-Marie stalked to the table's edge opposite the doctor, balling her fists. "What happens now? You show me how to pilot the spaceship?"

"Be quiet. Just look down."

Ann-Marie did as she was told. The tabletop was of smooth white Formica. "Oh, yes, this is very interesting," she mocked. The only things breaking the monotony were mirrors set at angles at either end. "I guess you really showed...." The words died in her throat as she caught sight of herself in the mirror.

A man stared back at her. "What the...?" She shot Beeks a questioning look and resumed her study, panic rising.

She pursed her lips.
The man pursed his lips.
She shook her head.
The man shook his head.
She slowly brought her hands up and felt her cheeks.
The man did the same.

"Oh, my God!" Her breath started coming faster. The man in the mirror was becoming flushed. Sweat popped out on his forehead. Ann-Marie wiped at her own forehead, and her hand came away wet. "What have you done to me?" She could hear the terror increasing in her voice. "Who is this?"

She clawed at her cheeks.

The man in the mirror bled.

He cried, too.

"No!" Ann-Marie yelled. "This is impossible! I am *me*!"

Beeks rounded the console and tried to pull Ann-Marie away. Ann-Marie threw her off violently. "Who are you?" she screamed at the reflection. She raised her fist and smashed the glass.

The madman stared back at her from each tiny shard.

She felt something bite her neck, and the head in the mirror jerked to one side. The man's face came closer very slowly. Or was she falling toward it? Things went black before she discovered which.

"What in the world is goin' on with you?"

Ann-Marie continued shuffling the hand she was dealt. "You gave me crap here. Did you rig this deck? I have nothing to put down."

"Don't sass foolish with me, child. You know I ain't talkin' about gin rummy." Evangelene took Ann-Marie's cards and threw them on the pile. "You look like a polecat that's been caught in a squall."

Ann-Marie laughed. *That* one was new. "Just prison pallor. The onset of rickets, maybe?"

Evangelene remained stoic. "The only bones gettin' soft are the ones in your head."

"Okay, okay. The parole board sent back another refusal. They won't consider an early release. It upset me."

"Enough to make you wake up screamin'?"

Ann-Marie stood reflexively. Last night's scream had been the first. She couldn't remember what had caused it (of course), but it was the most violent awakening she had ever had. "What do you mean?"

"I told you to cut the fool talk. They heard you over on

the next cell block, you were carryin' on so."

Ann-Marie paused in her pacing. "Carrying on?"

"Girl, you were moanin' and groanin' enough to wake the dead. The screamin' was only the finale. Woke me out of a sound sleep, it did. So I'd say you owe me an explanation."

It had started *before* she woke up? Acid soured her stomach and fury engulfed her. "You want an explanation? Fine." Her tone turned accusatory. "And I know you can't *bear* to hear my sob stories, so I wouldn't *imagine* inconveniencing you with a grand display of my petty character flaws. I'll make it short. There's something I can't remember; something that haunts my sleep but disappears when I wake up. And I know it has something to do with why I'm stuck in this rat hole." Ann-Marie felt tears running down her cheeks. The words scraping through her gritted teeth prevented Evangelene's attempts to speak. "And don't even *dare* to start a lecture. I don't know what it's like to live in your serene little world of burdens borne and beds slept in. It must be nice. But *forgive* me for being human. I'm bitter. I'm angry. And I feel like I'm going. . . ."

Ann-Marie stopped herself. She refused to let the word pass through her mind, much less her lips. She would never lay herself bare like that. Not in front of *anybody*.

There was a look in Evangelene's eyes that Ann-Marie had never seen before—fear. Only then did Ann-Marie realize she was looming over her friend. She leaned back against the sink, wiping at her cheeks and sniffling. "There's your explanation. And if you can't deal with it, it's *your* problem. There's the door."

Evangelene stood, but didn't leave. Instead, she hugged Ann-Marie. Ann-Marie tried to pull away at first, but at last surrendered, letting the tears come.

"Shush, child," Evangelene soothed, stroking Ann-Marie's hair. "What could be causin' you such pain? Why didn't you tell me earlier?"

"You said. . . ."

"I *know* what I said. But respectin' your elders can only go so far. Sometimes you just got to tell an old fool to shut up."

Ann-Marie pulled away, attempting composure. "What

difference would it have made? Everyone's got a story, right? You can try and analyze it all you want, but nothing can change it.'' She sighed and forced strength into her voice. ''Just forget about it.''

Evangelene cocked her head, hands on hips. ''Girl, you just knocked down part of the wall. Why you tryin' to build it back up again? You *don't* have to do it alone.''

Ann-Marie snorted. ''The hell you don't.''

''Even if I told you I might be able to help?''

''How?''

''Come to my cell tonight an hour before lights-out.''

''Are you telling me you can help me remember?''

''Maybe. There's a difference between tellin' and doin', of course, and come right down to it, the choice is yours. But to make it work, you're gonna have to give me somethin' I can see you never gave no one.''

''What?''

''Your trust.''

CHAPTER EIGHT

The cell was bedecked with candles. Ann-Marie's eyes were drawn to their dancing glow, her skin warmed with their heat, her nose filled with their odd scent.

Evangelene sat with her back to her neatly made bunk. She smiled at Ann-Marie and pointed to the space in front of her. The flickering points of light multiplied Ann-Marie's shadow into a thousand overlapping images that followed suit when she sat down.

"Where did you get all these candles?"

"Child, just because I don't have a radio like you doesn't mean I don't have connections. I've been in this place a long time. When I want somethin', as you can see"—Evangelene gestured slowly around the room—"I get it. And as *I* can see by the look in your eyes, you're confused and maybe a little scared, although I know you'd rather eat raw chitlins than say it out loud."

Ann-Marie wanted to ask what in the world chitlins were, but she was afraid her voice would betray her. Instead she just looked questioningly at Evangelene.

"The candles are more to get you relaxin' than anythin' else. You need to calm your mind, empty it of thoughts, so I'll explain what I can to put your wonderin' to rest.

"When I was a little girl, Momma used to take me to visit my grandma, who lived on a small island called Sapelo. It's one of the Gullah islands off the coast of Georgia."

"There are islands off the coast of Georgia?"

"You got cotton in your ears, girl? What did I just get done sayin'?" Evangelene waited for Ann-Marie's nod before continuing. "There's a whole mess of islands there. People call 'em the Low Country. Those that live there are called the Geechee. Momma and me, we'd take a boat over there a few times a year. That's when Grandma would teach me about Olorun; that was her name for God." Evangelene paused at Ann-Marie's questioning look. "Go ahead. Ask."

"It's just that I thought you were a Christian."

"I am. But my grandparents were brought to the island as slaves, come over from Africa. They had a different religion. Since the islands are so isolated, they were never made to give it up. And when freedom came after the Civil War, all the white folks left. The African religion took hold stronger than ever."

Evangelene chuckled. "Daddy would *never* come with us. He was a God-fearin' man, and he used to tell me not to listen to Grandma's crazy talk about such things. But Momma explained it so I could understand. Grandma simply saw God in a different way. To her, he had lotsa faces, not just the one Daddy believed in. The faces were called *orishas*, and each watched over a different thing."

"*Orishas*?" Ann-Marie wondered where this could possibly be going.

"That's right. I remember her tellin' me about seven or eight of 'em. There was Obatalá who created the world and humanity after Olorun created the universe. Ogún watches over workers, and war. Oshosi is the hunter, Oyá is the wind, and Oshún and Yemayá control the waters."

The strange names rolled off of Evangelene's tongue in a stranger accent. It was Southern, almost, but there was something else. "Obatalá." Ann-Marie had trouble reproducing the sound.

"Don't even try, girl. Unless you come from Africa, you probably won't get it. Grandma had a real peculiar way of talkin'. Momma, she called it Yoruba talk." Evangelene poked at a glob of soft wax collecting around a candle. "Grandma never took to English much, bein' separated from the mainland like she was. I could only understand her half the time and couldn't but talk a little."

"But you learned the names of these *orishas*?"

"They were easy. You could see 'em all around. One year a flood come up with the rains. Grandma said Oshún and Yemayá were havin' a squabble. Whenever it was a'blowin' and howlin' at night, Oyá was the one causin' the ruckus."

"How are nature gods going to help me?"

"I was gettin' to that, if you'd let me. There were others— those who looked after things on the inside, if you get my meanin'. Elegba was one of them. He stands at the crossroads between the world of the flesh and the world of the spirit. It's him that allows us to open up doors and travel down roads. His colors are red, white, and black." Evangelene said that with a weight that told Ann-Marie it was supposed to mean something. Her silence annoyed the older woman. "Are you blind as well as deaf?"

"Huh?" Ann-Marie looked around in the flickering brightness, finally focusing on the wax Evangelene was dipping a finger in. It was red. *All* of the candles sat in pools of red, white, or black wax. "You don't mean...."

"It's precisely what I mean. Elegba can help you."

"Evangelene, no. I don't believe in God, and even *I* know it would be blasphemy for you to do this."

"How can it be blasphemy? There's only *one* God. When Momma told me that Grandma saw many of His faces, she was only helpin' a fool girl understand somethin' a little more complicated. But since God is everywhere, why *can't* He show Himself in different ways to different people?"

"But you're talking about something primitive here. A religion practiced with ignorance of—"

"*Surely* you're not calling my grandma ignorant, or primitive? She was neither. Nor was she evil, as Daddy thought. That's why the more I think on it, the more I'm convinced Momma was as close to the truth as a bullfrog is to mud; it really *was* God's way of reachin' Grandma and her folk. And Grandma taught me how she would talk back."

Ann-Marie was still uneasy, but she wasn't going to argue. *This* was the big answer? She was positive it wouldn't work. "So what do we do?"

"We're already doin' it. It's called a *bembé*, a meetin' to connect us with the spirit world. The candles are an offerin' to Elegba, invitin' him to join us." She shot Ann-Marie an accusing look. "Your eyes are tellin' me you've already given up on him. That's why you need to trust me, like I told you. A bit of faith is all I ask. Some belief in somethin' other than what you can grab up in your hands."

"What do I do?"

"Surrender. Let down your barriers."

"But. . . ."

"Don't 'but' me, girl! Why are you so afraid? I'm tryin' to help you find truth. Instead, you'd rather keep kneelin' down to fear. Don't you realize that's more of a surrender than anything else?"

Ann-Marie drew a ragged breath. She didn't give a damn about fear *or* truth. She just wanted it all to go away. But she needed to know. More than anything, she need to know. "I'm ready."

Evangelene smiled. "Good. And I promise not to hurt you. Just close your eyes and relax. The world we're livin' in ain't the only one. If you want to travel the path of spirit, you need to open your heart. The answers you're lookin' for are there."

Evangelene started to chant rhythmically, keeping time on her knees. Ann-Marie listened closely, attempting to decipher the words. But it was hard to concentrate. The steady mutterings bore her tension away.

"That's right," Evangelene continued in the same meter. "Let it all go. Open your mind." She resumed the chant, adding subtle nuances to each repetition.

Ann-Marie felt her mind draw in on itself. Slowly, slowly.

The flickering afterimages on the insides of her eyelids were swallowed by an ever-brightening blue glow.

Limbless, breathless, swimming in the light.

A light of cold blue fear, of harsh white loneliness. Another (person?) passing by her, through her.

Where am I?

The thought echoed unanswered, its dying reverberation a menacing companion.

Distant lights moving closer (she moved closer to them?), larger, larger, coming together. A face. Black. Large eyes.

Sociological experiment... looks like you, talks like you....

Words spilling into each other, prying into (prying out of?) her thoughts, naming themselves.

Beeks.

Getting fainter, dissolving.

Alone. Insignificant.

Lights. Wrapping her wrists. Biting.

Trapped!

More lights. Coalescing. A harsh white table. Upending. Getting closer.

Run!

Closer. Closer. Coming from every direction.

No!

Harsher white on each end. Terrible whiteness. Searing.

Away!

A man's face. Kind face. Crying. Bleeding. Terrible face.

I am me!

Many faces. Twisted. Spinning away in every direction.

I am me! I am me! I am me! I am me!

"I am me!" Ann-Marie's head snapped back, connecting with the stone wall and forcing her eyes open. Sweat stung them, refracting a thousand dancing flames into a thousand more. She pulled up her soaked shirt, wiping vainly.

The memory jarred into place, filled the chasm, banished the old questions, brought new ones.

Evangelene rushed over to Ann-Marie. "Girl, you take more knocks to the head than a two-penny nail! How does it feel?"

Ann-Marie patted gently at the tender area. No blood stained her fingers. "Just a lump this time."

"What did you see?"

Ann-Marie got up dizzily and rushed out onto the block.

"Whoa-ho-ho! Look at you! You ladies having fun in there?"

"Fuck you, Tibor."

After Tibor ordered her out of Evangelene's cell, Ann-Marie went directly to her own. Now she was sitting on her bunk, her pad on her lap. Her pencil worked furiously, reconstructing the memory that was getting stronger with each line drawn.

A long face, largish nose, defined cheekbones, but not harsh. Deep hazel eyes.

Evangelene had followed her. She cast a shadow into the room. "What is it? What did you. . . ."

Ann-Marie held up a silencing hand. Round forehead. Brown hair pulling away from a white streak.

Sweat dripped onto the page, blurring an eye. She brushed at it in annoyance, smearing the image. "Damn!" She moved on to the chin, round just like so. . . .

It was finished. She held it up for Evangelene's inspection. "He's the key."

Evangelene took the pad and studied the face. "Who is he?"

"The man of my dreams. The man in the mirror."

Evangelene looked at her in confusion. Ann-Marie overturned the mattress, pulling out the pictures and lining them on the floor. "Look. Look at these." She pointed to the image that brought her the most fear. "He's the shadow man." She indicated Beeks, the table, the bunk. "She was there, and that was there, and that was there." She rambled on, not caring that she sounded like a demented Dorothy returning from a place far stranger than Oz.

"Where, child? Where?"

"I don't know. But I remember. I *remember*." She threw her arms around Evangelene and hugged her fiercely. "Thank you. You did it. Thank you." Tears leaked out of the corners of her eyes, and she felt ridiculous.

Crying from too much happiness was something that happened only in pages of books. She threw her head back and laughed. "I don't think I can explain it. Not now, anyway."

"But. . . ." Evangelene sighed. "Okay, okay. It's gettin'

close to lights-out, anyhow." She raised an admonishing finger. "But I want to hear all about it come mornin'."

"I swear to God."

"Don't blaspheme where I can hear it, child. I'll pray for you before I go to sleep. If I ever *get* to sleep, with all those candles that need pickin' up...."

Evangelene's voice trailed off as she left the cell. Ann-Marie softly swung the door shut and returned to the pictures lining the floor. They finally made sense.

Yet they were more confusing than ever.

This went far beyond an FBI sting. Where had the strange reflection come from?

She shivered. As good as it was to remember, she felt more spooked than ever. What the hell kind of experiment had she gotten caught up in?

Not sociology. It was something more, something they hadn't told her. Or *couldn't* tell her.

"Why me?" she asked the image of the mirror man. "Why did you put me in here? How?" She clutched the paper hard enough to tear the edges. "Who are you?" Anger contorted the question into a strangled squeal; frustration cracked her mind's edges.

Who who who?

Control! You have your answers.

But what good are they?

It's a beginning.

It was supposed to be an ending.

Be strong!

Yes. Strength. A beginning. She could not give in now. *Would* not.

"You're mine now. Did you really think you had the right to screw with me and get away with it?" The page quivered in her trembling fingers. "*Did* you?" The charcoal face remained mute, mocking.

CHAPTER NINE

The sun was sliding into evening, bathing the club in twilight. Sam walked around the room, studying the framed posters lining the walls.

Their distinctive style—colorful artwork, garish lettering, almost no photos—reminded Sam of Saturday afternoons in Elk Ridge when Tom would take him to the theater to catch the creature features. They were some pretty terrible movies, always more funny than scary, and these posters were just like those in Elk Ridge.

Well, maybe not *exactly* like them. If they were, his mother would have forbidden him even to walk *past* the theater, much less go in.

His favorite so far was *Satan's Seductress*. A rather supple cartoon Kat lounged to one side, sheer black negligee falling dangerously from one shoulder. Her face was cloaked in shadow, and blood dripped from the corners of her mouth. *One Kiss Is All She Asks* . . . read a line toward the bottom. A montage of smaller pictures clustered underneath depicted men in various states of terror. Sam thought one of them was Richard.

The posters for *Decadent Debs* and *Sorority Bloodbath* tied for second favorite. Of course, all the posters suggested much more than they actually showed, but that was what made them so effective. They made him want to see the movies.

"They're a lost art." Richard was suddenly beside him. The man's movements were as quiet as his manner. "That was one of the few posters I was actually featured on." He pointed to *Satan's Seductress*.

"That *is* you."

"Yeah, about twenty-five years and thirty pounds ago." Richard chuckled. "Maybe this will help." He popped his eyes out and dropped his jaw. Sam was amazed at the transformation; Richard looked like he was on the verge of insanity. An instant later he was back to normal, grinning slightly.

"You haven't lost your touch."

"That's nothing. You should have seen me when I had more hair." Richard poked a finger at his scalp. "I could make it stand on end."

Sam laughed. "Quite a trick."

"It was something, all right. Actors these days would be lost without all the slick effects and fancy lighting. Movies just don't have the *feel* anymore."

"Is this collection an homage to that?"

"Not really." Richard scanned the room. "It's more like my family album."

"Huh?"

"See that one?" Richard pointed to *Invasion of the Flaming Leeches*. "That's when Kat and I first met. Our first offscreen kiss was on the set of *Mutant's Revenge*. *Murderous Mandroid* was made just after our honeymoon." He indicated each poster in turn. "All of them mark different times in our life, in our youth."

"That's really . . . romantic." Despite the oddities, Sam was beginning to like this man more and more. His retiring ways were a good counterbalance to Kat's flamboyance; it was a possibility he had overlooked in preparing for the impending confrontation. "Would you consider speaking to the council tonight?"

"Me? I'm just a bit player. Kat's the showgal."

"And she's ready to ram it down their throats," Sam said. "She's worked herself up so much that I'm afraid she's gonna forget about her speech and just tear Barrenger's throat out."

"That's my scream queen," Richard said proudly. "She'll make 'em stand up and take notice."

"Well, that's only part of what we want. They *also* need to sit down and really listen. You can make them do that."

"What would I say?"

"Just speak from the heart. You seem to have a knack for it."

Richard's nod was reluctant. "I *do* have a stake in it all, I suppose. But if I come away from it looking like that"—he pointed at his poster double—"I'll come looking for you."

Sam nodded and smiled. "Deal."

Richard departed quietly and was soon in his office, leaving Sam alone to guess at the odds. Had he turned things in their favor or had he ruined everything? There was no way to tell, because he hadn't heard a word from Al since the night of the Observer's humiliating exit.

That in itself was more troubling than he could muster the strength to think of at the moment. Four days with no odds, no glaring neckties, no lecherous jokes. Al had stayed away from a gaggle of mud wrestlers for *four whole days*. That really put things in perspective; something had to be *seriously* wrong with Ziggy.

His worry was what had gotten him to the club so early. More was riding on tonight's showdown than the future of the Kit Kat Bar. If all went well, he would probably Leap.

What would happen then? If the Project hadn't contacted him for this long, would they even be able to reestablish the link after he Leaped out?

Don't do your job and you might be stuck here for the rest of your life. Do your job and they might lose you for good. He was feeling sick.

Kat barreled through the front door. Her burgundy suit, a

true creature of 1980s big business, was offset by a floral scarf. Her hair was gathered into a ponytail, and her makeup was minimal. The contrast to her usual appearance was so striking that Sam didn't recognize her until she spoke. "Anyone else here yet, Candy?"

"Just me and Richard." Sam let out a low whistle. "Boy, you clean up real good. That ought to knock Barrenger a little off balance."

"On the contrary." Kat fidgeted with the scarf and readjusted her lapels. "I think this is the one time in my life when I *can't* rely on my looks to get by."

"Don't worry. The girls know what to say. You've been drilling us all week." So much so that the club had been closed since the night of the rally. *That* was definitely a plus, since it meant he didn't have to wrestle. But Kat's attitude was contagious, putting them all on edge. He anticipated her next comment and raised a finger to forestall her. "*And* I brought a copy of each statement. They're on the bar. No one will forget her lines."

That seemed to appease Kat. She joined Richard in the office as the other girls started to arrive. They were all a bundle of nerves and anticipation, even the ones coming to show moral support and solidarity.

Tawny was one of the last to arrive. Since she wasn't speaking, Sam really hadn't seen much of her. Steeped in preparations, and with no Observer to remind him, he had almost forgotten about the altercation a few nights ago. He needed to get moving.

"I didn't think you'd show."

Tawny smiled. "Why? My paycheck is riding on this, too."

"It's just that I haven't seen you around much the last few days. Been busy with Kyle?"

"Real subtle." Tawny began to walk away. "Can't you just leave it alone?"

Sam grabbed her shoulder and spun her around. "Can't *you* face the fact that you can't walk away from this?"

Tawny shrugged him off. "That's the last time you touch me outside the ring without getting a broken arm, understand?"

In that case, I hope it's the last time I touch you. Instead of expressing the thought, Sam nodded and said, "But will you at least hear me out? After what happened, I promised Kyle I'd talk to you."

Tawny folded her arms.

"Don't you think you should go home to Kentucky for a little while and try to straighten things out? I'm sure your father misses you, and Kyle's certainly concerned."

Tawny threw her head back in a bitter laugh. "Oh, he's concerned, all right. *Kentucky.* He's concerned that he'll have to do everything himself. He's concerned that it might interfere with his precious *career.*"

"Don't like Kentucky, or is it journalism in general you're against?"

"Both. Kentucky isn't home. It's only the latest rung on Kyle's ladder to success."

"Huh?" He seemed to be saying that a lot lately. Too many variables were slipping him up. Where the *hell* was Al?

"Don't you know that's how it works? You get your degree, and then you work for minimum wage at some dinky station in some shitty little town in the middle of nowhere. A year goes by, and you move to another shitty little town, only this one's in a slightly bigger market and they've tacked benefits onto your slave wages. In a few years you do it again. And again."

"Everyone needs to start at the bottom," Sam said. "You can't fault him for that."

"I don't. I hope he makes it to New York, or wherever journalist nirvana is. But I won't give up my life to do it with him."

"Why would anyone expect you to?"

Tawny shook her head. "It was some stupid promise that he made to Mom just before she went—my mom passed away."

"So I gathered. I'm sorry."

"Yeah, well, so am I. He told her he would keep the family together, take care of Dad and me. But his job takes up all of his time."

"So he has to work hard," Sam said. Did she realize how

childish she sounded? "I don't see how that can drive you to run away from it all."

"He expects us to come *with* him. He wants his career so bad, but he feels guilty leaving us behind. So it makes him feel better to drag us all over the country whenever a new job comes along."

So much for snap judgments. "Doesn't your dad have anything to say about this?"

Tawny let out a laugh more bitter than the last. "That chameleon hasn't had anything to say about anything since Mom died. The night the hospital called and told us, he hung up the phone and just started walking around the kitchen in circles, mumbling to himself. I had to do everything. I didn't mind it then. He was in a lot of pain and Kyle was so busy with school that he couldn't make it home until the funeral. I was needed, and I did what I had to."

"Why did you call your dad a chameleon?"

"Because I kept expecting him to get better, to snap out of it. But he didn't. Sure, he put on a good show for Kyle at the funeral. Couldn't upset the Golden Boy with exams and graduation coming up. So I continued to do it all—shop, keep house, make mortgage payments—because my father just sat in his chair day after day, pining away."

"Why didn't you go to a counselor?"

"Who the hell does things like that? I was too busy, anyway. And I expected it to get better when Kyle was finished with school."

"It didn't, did it?" The story was sounding all too familiar to Sam.

"Worse. Kyle was working like a dog, sometimes sixteen hours. The only time we saw him was on TV and when he'd come home and collapse. Oh, Dad would perk up then. 'Giving 'em hell, son? Great story tonight.' It was all I could do not to throw up, the way he went on."

"So to Kyle, nothing was wrong." *Very* familiar.

"Only with me. I complained too much. I had to face my responsibilities like an adult. There was nothing wrong with Dad that *he* could see. The truth is, he refused to see it. It would interfere too much with work. After he started talking about moving us all to Kentucky for this great new job, I bolted. I'm not going to spend the rest of my life chained to one man who won't help himself and another who's using me to appease a dead woman."

Tawny took a breath, letting the anger loose. "It's a real bitch, ain't it? I chose to have my own life. And if you see Kyle again, let him know. I haven't spoken to him since the other night." She moved off.

Sam watched her go. If her point didn't hit so painfully close to home, he might have thought of a response.

"Daddy's dead, Sam. Tommy's dead. Mom can't run the farm by herself."

"But you can't just sell it. Mom's spent her whole life here. It'd kill her to leave."

"Well, then, who's going to help her? You? You spend all your time going from M.I.T. to Washington and back, doing whatever it is you do. If you took your nose out of a book for more time than it takes to eat Christmas dinner, you'd know what it's like to watch her live day in and day out with the ghosts in this big, empty house."

"But she seems fine."

"Do you really think she'd let you believe otherwise? Jim is my husband, and I have to go where he does. And since I won't leave Mommy here to rot, there's no other choice. She's coming to Hawaii with us."

Katie had been right, of course. Cold comfort. *First you're not there for Dad, then you continue the vanishing act when Mom needs you.* How could he possibly fault Tawny?

He wiped at the corners of his eyes, forcing the memories away. How come Swiss-cheesing never worked in his favor?

Personal hells aside, he still needed to reconcile Tawny with her brother. Ziggy had said saving the club was his job, but Ziggy had been wrong before. Without Al (and maybe even with him, considering the shape Ziggy seemed to be in) there was no way for him to know for sure.

But how are you going to do it for them when you couldn't do it for yourself?

The council meeting had been going for almost two hours, and the end seemed nowhere in sight.

Town Hall was hosting *the* event of the evening. Sam had

marveled at the crowds gathered outside when they arrived. Picketers carrying RPL signs contended with another group carrying signs reading "Take Our Liberties? No Way!" and "Keep Wilson Free!" His favorite said "A Vote for the Mud Queens Is a Vote for Our Children."

The media had also made a good showing, flashbulbs lighting the night like strobe lights.

The agenda for the meeting had been in the town's paper; aside from Kat's hearing, the only other business on the docket consisted of a rezoning proposal and a broken curb on Main Street. Sam had in no way expected a reaction like this. The townspeople hadn't let him down after all.

Many had prepared comments of their own, presented between his statements and those of the women from the club, all chiding the council for even considering such a measure, and Barrenger in particular for spearheading it. Others showed full support. It was truly a town divided.

"Is there anyone else who wishes to speak to this order of business before we vote?" Sam held his breath. They had agreed that Kat would be the last one up.

It was a trick he had learned from Al when faced with Senate committee funding hearings: save the big guns for last so the blast will still be echoing when it comes time to vote.

"Madam Mayor, members of the council, I would like to address the situation if I may." Kat's tone was more respectful than he had thought possible. Reporters vying for lead story footage and front-page shots made the already crowded room seem more so.

The seven council members looked up in anticipation, but whether it was for what Kat had to say or because her presence signaled an end to the long debate, Sam couldn't tell.

The mayor nodded. "I suspected you would."

"Thank you, Mayor McClough. As you all know, I am Kathy Scherber-Danson, owner of the establishment that seems to have caused such a commotion lately. An oddity in its history, since I have always tried to run a quiet establishment.

"We have heard many people speak here tonight, town residents and my employees alike, basing their arguments on

First Amendment freedoms. But, as I'm sure the esteemed members of the press joining us tonight can attest"—she acknowledged the throng surrounding her—"those arguments are open to interpretation. So I'll just stick to the facts.

"Fact: I am well aware of the stigma some people attach to the brand of entertainment the Kit Kat Bar offers. To head off the concerns of my fellow small-business owners and townspeople, I chose a location on the outskirts of the town's business loop, far from the center of commerce and residential areas. And I have strived to make it a respectable establishment—"

Barrenger's snort stopped Kat's speech. She remained silent for a moment, breathing deeply. Sam realized how tightly she was reining in her emotions. He prayed she could hold on for a little while longer.

She cleared her throat and fixed her stare on Barrenger. "A *respectable* establishment offering the adult community of Wilson a place to congregate and enjoy themselves.

"Fact: We've never been cited for admitting minors, and to try and ensure that they don't even know about us, we've never called attention to ourselves." Kat's slight pause here had been rehearsed over and over again. "Unless, of course, you count the drought two years ago, when we let the mud in the ring dry up in a show of solidarity." The expected laughs came. Even the mayor grinned and covered her mouth.

"Lately, however, a lot of attention has been thrown our way," Kat continued, "by a group calling itself the Rural Purity League. Leading this faction is council member Rex Barrenger. It seems that Mr. Barrenger has taken it upon himself to set the standards of decency for the people of this town. Surely, there are many that agree with him. But I think tonight's meeting proves that many people also support the right to make their own decisions, to come to their own conclusions about what constitutes purity.

"Fact: Mr. Barrenger's antics have called more attention to my little bar in the last few weeks than I have in the last four years. Fact: Mr. Barrenger and his group have disrupted my establishment, have trespassed on private property with *peaceful* protests, and have committed acts of vandalism potentially dangerous to my patrons and my employees, as the police record will attest. Fact: Mr. Barrenger has targeted my club in a very public witch-hunt and has used his so-called crusade as a bully pulpit to launch his mayoral campaign.'' Kat's voice had risen an octave by the last sentence, and she paused to breathe once again.

Barrenger was a picture of serenity. "No reason to get excited, Ms. Danson. It is this council's responsibility to address the concerns of the townspeople and ensure everyone has their say. *Whoever* they may be."

"Mr. Barrenger, I ask you to hold your comments until the speaker is finished." Mayor McClough addressed Kat. "Was there anything else, Ms. Danson?"

Kat nodded. "Just one more thing. I know you will retire at the end of your term and that Mr. Barrenger will likely take on your responsibilities. That is why I forced his hand and called for a vote tonight. My business license has *your* signature on it. My club operates under *your* authority. It is your leadership and the action of this council that have made Wilson a progressive, prosperous town. I urge you *all* to consider the legacy you wish to leave. Please don't let it be one of bigotry, fear, and ignorance. Thank you."

Many members of the crowd stood and cheered. Sam was among them. It had gone as well as they could have hoped. But was it enough? He didn't feel a Leap coming on.

"Order, please." The mayor rapped her gavel, silencing the crowd. "Mr. Barrenger, I believe you wished to speak?"

"I would just like to agree with Ms. Danson on one point. This council *has* worked hard to make Wilson the place it has become. But we have built upon a foundation of positive ideals. I am merely striving to uphold those ideals. Before we vote, *I* urge my fellow council members to remember the words of Winston Churchill: 'Do not let spacious plans for a new world divert your energies from saving what is left of the old.' If Ms. Danson chooses to label these energies as a witch-hunt, that's her affair. I have confidence that the rest of you will see things in their proper light."

This time it was Barrenger's contingent that cheered. Sam found himself admiring the man's persuasiveness even as he cursed it.

"If that is the end of the discussion"—Mayor McClough's voice rose above the crowd—"I would like to call a...." A slight throat-clearing stopped her, and Sam turned to see Richard standing. "You wish to address the council, sir?"

Richard nodded and approached the microphone. "I'm Richard Danson, co-owner of the Kit Kat Bar. I hadn't planned to speak here tonight, but Mr. Barrenger's quote reminded me of another string of words I heard a long time ago." He took a breath, and his voice issued forth in the same announcer's tone that Sam had first heard from him. "'He is an unprincipled and degenerate liar—but with a tremendous audience both in newspapers and on the airwaves. A man who has been able to fool vast numbers of people with his fake piety and false loyalty.' Now, I don't recall who those words were aimed at, but I remember vividly when Senator McCarthy said them. He was another one who duped a lot of us way back when. The point I'm trying to make is that wrapping rotten intentions in fancy words, no matter how fitting those words may seem, doesn't make a bad idea any better. It just makes those who have the bad idea more dangerous."

The hall ruminated in silence as Richard took his seat. If there was a better way to get the last word in, Sam couldn't think of it.

Mayor McClough spoke into the lull. "Time to vote, ladies and gentlemen. Decision will be by secret ballot, as agreed. We have heard many constituents speak persuasively tonight. I ask you to make your decisions based on their wishes, not personal agendas."

He couldn't say for sure, but Sam thought McClough was in their corner. The hall was as quiet as a monastery while the council members made their decisions and passed them to the mayor. The tension made his teeth itch. The worst part of it was that he should already know the outcome. If he was ever able to get his hands on Al again....

"Ms. Meyers, please verify these results." The clerk moved to the mayor's side, nodding at the papers gathered. "So noted," McClough continued. "The council hereby revokes the business license of the Kit Kat Bar, four votes for, three against. The ruling is effective immediately.'' She banged her gavel to silence the cheers coming from the Rural Purity League's members. ''Ms. Danson, failure to act accordingly will result in fines, imprisonment, or both.'' Her tone was official, but her eyes told a different story.

Kat remained silent as she and Richard left. The women from the club followed them. Sam watched in detached shock. He had followed all of his instincts, but they had betrayed him.

What am I gonna do now? What will happen to Sarah and the rest of them? Please, please, please, *God, bring Al back.*

The hoots coming from the other side of the room penetrated the haze. He was suddenly furious. ''You think you've won, don't you?'' Sam looked from the crowd to Barrenger. How he wanted to punch the smug smirk off the man's face! ''Truth is, we all lost. Nobody should be proud of what went on here tonight.'' He made his way into the cool evening, shoving reporters out of his way.

Kyle watched from his perch in the back of the hall as blondie stalked out. He would bet his benefits package that her sound bite would make the opening of every newscast in town. He could almost hear all of the producers within thirty miles yelling ''Eureka!'' when that little nugget made it onto tape.

The hall seemed much larger after the exodus. The council's comments took on a godlike quality, echoing off the vaulted ceiling. Barrenger certainly looked the part, sitting back in his oversized chair and staring down his nose at his fellow council members. He led the discussion as much as he could without being overtly controlling, Kyle noticed. The reporter had a good idea why.

If Nancy's tips had been right, and he had no reason for doubt, the real story was yet to come.

''It'll work just as well to drown our sorrows as it would have to celebrate with,'' Kat said, popping open a bottle of champagne.

The cork's flight to freedom was cut short by the roof. It

ricocheted and landed at Sam's feet. "I know just how you feel," he murmured as he picked it up.

"Come, now, ladies. Drink up. None of us have to get up for work in the morning, after all."

Some of the girls let out a halfhearted laugh as the champagne glasses were passed around. Sam declined his and threw the cork at the streamers hanging from the ceiling. He had been so *sure*. What now? The Kyle and Tawny situation still needed to be resolved, but what if that wasn't it either?

"Let's go, Candy," Kat cajoled. "We may have lost, but like you said, it was on our own terms. We couldn't ask for anything more. I want to thank you for that."

Glasses were raised, and all eyes turned to Sam. What could he say? "I guess that will have to be good enough." *Even if it means living the rest of my life as Candy Apple.* "What happens next?"

"Next, you all get two months' severance pay and we go our separate ways. Richard and I discussed this possibility earlier, and we decided not to stay if it came down to this. It's just a matter of principle now. Screw this town." Kat's plastic glass rapped off her husband's, and she quaffed what was left.

Sam's chest suddenly felt as if a small mammal were trying to burrow out of it. His instincts hadn't been right before, but he somehow knew that if Kat acted on her last remark, he would never Leap. His mind whirled in panic, discarding one idea after another. "What then?"

"I figure out what to do with the rest of my life."

"Any thoughts on what?"

Kat shrugged. "Open up a new place in a new town, maybe. Crusade to save the drive-in. Who knows?"

Sam finally found an idea solid enough to grasp, but he doubted it would keep him afloat for long.

Stupid, yes. But what the hell do I have to lose?

"You talked about sticking to your principles before," Sam said. "How far are you prepared to go to do that?"

Kat shrugged. "As far as I'm able, I guess, until there's nothing left I can do. I think I reached that point earlier tonight."

"Maybe so," Sam said, "but then again, maybe not. You may just have to set your sights a little higher."

"What do you mean?"

"How does Mayor Scherber-Danson sound?"

CHAPTER TEN

Ann-Marie had been wearing her best smile so long it was beginning to hurt.

Control. Just a while longer.

The men and women assembled at the long table in front of her shuffled their papers, returning the bulky sheaves to order. The man seated in the center looked down one side, then the other, before nodding and returning his attention to her.

"It is the decision of this committee, Ms. Renerie, that early release be denied."

Denied again. Calm.

"May I ask why, Mr. Martinek?" *Stupid little prick!* "As you can see, my record has been exemplary. I haven't gotten a charge sheet in years. I have—"

"All you have *done*, Ms. Renerie, is behave in the manner expected of you, or any other inmate in this facility."

Calm calm calm. Keep smiling.

"But it's only *one* year. Surely my good behavior counts for something?"

"We are well aware of the extent of your sentence, Ms. Renerie, as we are of your record." The weasel pushed his glasses higher onto the bridge of his nose and steepled his hands in front of him. "But to put it bluntly, this committee feels you are getting off lightly as it is. You plea-bargained your way into a state facility for a federal offense. Your sentence was reduced for your initial cooperation...."

Not my *cooperation.*

"... with the authorities in helping to apprehend your partners in crime. We would be remiss in reducing it further. You will not serve a day less than the ten years the judge decreed." He motioned to the guard, signaling the end of the conversation.

Another time, an earlier time, Ann-Marie would have been seething at the casual, matter-of-fact manner Martinek used when decreeing her fate. But she had experienced it enough by now to know further argument would be pointless. So the false smile remained as she let the guard escort her out of the room.

One more year. One more year. One more year.

The thought echoed with each stride.

It wasn't really the specter of another year on the inside that consumed her. That was old hat at this point; she could do it in her sleep.

What it *really* meant was another year where she couldn't get to *him*. One more year with him getting off scot-free. And all the while those prize morons with their polyester ties and ten-dollar shoes kept going on and on about justice as they saw it. She could choke on the irony.

Which was the greater injustice?

"I'll take her from here, Joey."

The guard nodded and she was handed off to Tibor. "Good afternoon, Mr. Tibor," she said as pleasantly as she could.

"My, my, Renerie," Tibor chortled, "aren't *we* polite? What, did the parole board scare some civility into you? Or was it your girlfriend, Robbins? It's a shame. I miss the little chats we used to have. But I'm warning you, these May–December flings usually end pretty badly. Don't let her break your heart."

Control.

Ann-Marie willed herself to remain silent. She wanted to smack the fat bastard in his moronic grin, but she would not do *anything* to risk tacking more time onto her sentence.

He means nothing. You have bigger game to hunt.

A few piercing bells and two stupid comments later, she was alone in her cell. A picture of the bird lady greeted her as she entered.

The rendering was appropriate, considering what the woman had called herself. Her large eyes rested atop a hooked beak, opened in midsquawk. The plumage cresting her forehead completed the image.

"Polly want a cracker?" Ann-Marie turned her laughter to the picture taped next to bird woman's. It was the latest addition, and it was her favorite.

Bloody sockets were all that were left of Mirror Man's eyes. He was straining against the straps that held him to the bed as multiple bird women gouged chunks of flesh from his body.

"You've bought yourself another year, Mr. Mad Scientist." She turned, addressing the other pictures papering her cell. Drawing him had become as mindless as doodling, she had done it so often. Each sketch was a reminder. Each meant that much more power. "Well, enjoy yourself. My turn will come soon enough."

"Who you talkin' to?"

Ann-Marie stopped her circuit when Evangelene crossed her vision. Her neighbor stood in the doorway, looking uneasily at the images tacked all over. "No one. Come in."

Evangelene made no move to enter. "How did the hearin' go?"

"The usual. You're stuck with me for another year. How'd the physical go?"

"'Bout how you'd expect. Like I need 'em tellin' me I got tired old bones." Evangelene's remark fell flat with the apprehension in her tone.

"Just as long as *I'm* stuck with *you* for another year." Ann-Marie waved her friend forward. "Will you come *in* already?"

"Why don't we play in my cell instead?"

"Why?"

"I just thought it might be a nice change.... Did you see the new memo Ward posted last—"

"Don't change the subject. What's the problem?"

"Truth is...." Evangelene hesitated.

"Geez! You look like you're in pain. Will you just say what's on your mind?"

"Okay." Evangelene stepped in and closed the cell door partway. "But remember, I say this as a concerned friend. Truth is, you're startin' to frighten me. Ever since that night we called up Elegba, you've been gettin' stranger an' stranger."

"*Wooo-oooo-ooo*!" Ann-Marie waved her hands above her head and stomped toward Evangelene. "You've found me out. The evil nature demon has possessed me. Join us, Evangelene. *Joooinnn uuusss.*" Ann-Marie trailed off into a laugh. Evangelene didn't join in.

"I *mean* it, girl. This ain't no joke. Sometimes you seem positively touched."

"How so?"

"Well, for one, you were just spinnin' yourself dizzy talkin' to walls." Her eyes fell on the image Ann-Marie had been studying. "And look at this! It's disgustin'. It's just downright disturbin' is what it is."

Ann-Marie moved in front of the picture instinctively. Protectively. "I'm only doing what you told me to. Finding a focus. Taking it day by day."

"I said to find somethin' constructive, child. Not *this*." She threw her arms out from her sides.

"*This* is what I've been looking for all these years! I finally know the reason I'm here. It's like I told you."

"Yes. The blue room and the mirror. But I worry that you're not rememberin' properly. Somethin' certainly happened that crawled underneath your skin. I don't doubt that. But strange folk switchin' bodies with you? It's just not...." Evangelene stopped herself, lowering her eyes.

"Go ahead, finish! It's not *sane*, you were gonna say!"

"I don't mean it that way. Please believe me. All I mean is maybe you should talk to someone about it. Someone besides me. Just look around you, girl!" She tore down one of the many pictures. "This ain't healthy. You're obsessed with someone that could be nothin' more than a haunt prowlin' around in your head."

"That's not true! Elegba showed me, like you promised."

Ann-Marie crossed her arms. "You're calling your god a liar now?"

Evangelene shook her head. "Not at all. How dare you even suggest it? But sometimes, he doesn't tell us things straight out. You got one bit of knowledge, but are you usin' it in the right way? As my daddy would say, 'The difference between a wise man and a fool ain't the amount of knowledge they have, but how they use what they got.' "

"Well, it's the only bit I have, and I'm using it to pull myself up. To lead me out. If that seems strange to you, then maybe *you're* the crazy one. There's a saying where I come from, too. 'Knowledge is power.' For the first time in years, I feel like I'm back in control. And I'm certainly not going to any prison shrink when I'm this close to getting out. And if that doesn't jibe with the notions of your dead daddy, then to hell with him. And to hell with you."

"Ann-Marie!"

Ann-Marie could hear the hurt in her friend's voice, but she had crossed a line. A step back would be weakness. "I mean it, *child*. Is there any pleasing you? All along, I've done everything you've told me to. I *trusted* you. And what happens? You come in here and call me crazy. Maybe the problem isn't with me at all. Maybe *you're* the one who's crazy. You *have* to keep that damn smile plastered on your face, keep up the cheery fucking attitude no matter what, because you have nothing left to look forward to. You're gonna die in this shit hole. But *I'm* going back to the world, back to normal life—"

"I know you're upset," Evangelene said, anger shading her words, "but that's *no* reason to act such a—"

"Save it! Would you please just spare me for once? Let me tell you a saying *my* father was fond of." Ann-Marie pulled the cell door open and pushed the other woman out. "It was 'Tell your story walking.' Now get the hell out of my sight." She slammed the door on Evangelene's next comment.

You don't need her.
Right! To hell with her!
You don't need anybody.
Nobody! Except....
Ann-Marie caressed the nearest image of the Mirror Man.

The morning bell shattered Ann-Marie's uneasy sleep. Her entire body felt clammy, and her mouth tasted like ashes.

Fuzzy images melted away as she shook off her grogginess. The same images were responsible for her restless night, and for once they had nothing to do with her strange experience.

Evangelene was at the root of them. And emptiness. The events of the previous day crushed down upon her.

"Stupid, stupid, stupid...." Her open palm met her forehead repeatedly as she let out a string of curses, damning herself. What had possessed her? It was like another person had taken up residence in her head and pulled the strings controlling her mouth.

What have I done?

No voice rose to answer.

But the night's twisted images were answer enough. She had alienated the one person to whom she owed everything.

She was on her feet and out on the block in the next instant. She had to apologize, to make things right again. It wasn't a question of weakness; it was one of sanity.

Evangelene wasn't with the other women forming ranks for the march to breakfast. Her cell door stood slightly ajar, automatically unlocked with the ringing of the bell. She was still inside, getting ready. Even better. No one else had to see Ann-Marie eat crow.

She pushed the door open, but the cell was quiet. Evangelene lay on her cot, face to the wall, head propped on her hand. She didn't turn toward Ann-Marie. For her not to show even the most rudimentary courtesy, Ann-Marie knew her friend had to be extremely angry.

Ann-Marie closed the door behind her. "Evangelene, I'm sorry. I didn't mean it. I don't know what came over me. It was just the pressure, the strain of knowing, but still *not* knowing. It's starting to take a toll. But I never meant to hurt you. Can you forgive me?"

Nothing.

"Evangelene, please. I know they sound like excuses to you. But you're the only real friend I've ever had. I need that. I need *you*. I can't bear the thought of what it would have been like in here without you to help me."

Still nothing. Ann-Marie sat on the edge of the bunk, having trouble getting the words around the lump forming in her windpipe. "*Please* talk to me. Even if it's just to say you hate me." She gently shook her friend's shoulder. "I need to hear your. . . ."

The words caught in Ann-Marie's throat as Evangelene's body rolled toward her, stiff, complexion ashen, arm still crooked at the elbow, balled hand still pressed against her cheek.

"Evangelene?" She shook harder. "Evangelene, wake up!" She began throttling her friend's corpse. "Wake up, wake up, wake up." Each syllable accompanied a push down, a pull up, both words and movements becoming more frantic. She let out a wail and clutched Evangelene to her. "No! Please, God, no!"

The cell door burst open. "For Christ's sake, what's all the shrieking going on in here?" Ward barreled her way into the cramped quarters, Tibor only a half-step behind. Both stopped short.

Ann-Marie rocked Evangelene back and forth, crying softly into her hair. "God, bless her. Watch over her. Please, God, please." Ward laid a gentle hand on her shoulder.

"She was almost eighty, Ann-Marie. It was just a matter of time."

The sincere tone was lost to Ann-Marie. Rock back and forth, back and forth. "You're with Thomas now. Your Thomas, your baby. Bring her home, Thomas. Bring her. . . ." The words dissolved into sobs.

Tibor pried her arms away and led her back to her cell, closing the door gently as he left. She didn't know how long she lay on her bed, weeping, before she dragged herself to her feet.

She splashed some cool water into her swollen eyes,

washed the saltiness from her cheeks, and prayed that Evangelene could hear her and forgive.

Please, God, let her hear me.

Stop kidding yourself.

Shut up! Wherever she is, she knows I'm sorry.

She's dead, and she died hating you. She died cursing your name.

No! That's not true!

Weakling. Idiot. Face it, you killed her.

Ann-Marie squeezed her hands to her ears, frustration rising with the voices.

No! Be quiet!

Murderer.

She cried aloud, and her eyes fell on the sketches on the wall. *He* was there, staring back at her from every direction. Silent. Menacing.

"You! This is all your fault! You killed her!" Ann-Marie attacked the face in front of her, shredding it with her nails, shredding her fingertips on the stone wall beneath. "You're the murderer! Not me! You!"

Blood smeared the lead. Red haze obscured her sight. Red hate filled the voices.

Strength.

Power.

Will.

Vengeance, child, vengeance.

CHAPTER ELEVEN

Tina's lower half dangled out of the service conduit, short skirt straining around her thighs. Al leaned against the opposite wall, puffed contentedly on his cigar, and enjoyed the way her stockings fitted her calves. For the first time in days, he forgot to be worried.

Four days of hunting. Four days of diagnostics. Four days of feeling useless, peering over the rim of his coffee mug as technicians gave Ziggy the workout of her life.

Tina squirmed backward, high-heeled feet stretching for the floor. Al took a last longing look before pulling her out. He rubbed at the smudge of grease decorating the end of her nose as she brushed her tangled red locks out of her eyes. "Well?"

"Nothing. I've been through every square inch of this mainframe. The hardware's, like, not the problem."

"Well, Gooshie says it's not the software either." He had said it enough times, as a matter of fact, to give Al fond remembrances of the oxygen mask in his jet. The stink of the man's breath lingered in his nostrils and clung to his clothes; he would probably have to burn one of his favorite shirts if the third wash didn't do the trick. "So where does that leave us? And Sam?"

Tina pursed her lower lip in thought, gloss catching the reflection of Al's blinking lapel pin. He hoped it would be a *long* thought.

"Maybe we're, like, looking too hard. The problem could be simpler than we think."

"How so?"

"Well, we've eliminated the possibilities of structural damage or program failure, right? So it *has* to be something else." Her face brightened in a way that made Al's insides burn, and she took off down the corridor. "Come on."

Al trailed behind, wondering how she could manage that speed in those shoes. By the time he caught up to her in the Control Room, she was kicking a pair of white-clad legs sticking out from beneath the main console.

"Gooshie, come out here. I think I got it. You too, Fuller. Front and center."

Samantha joined them as the head programmer pulled himself to a sitting position and rubbed his shins. "Simple words would have sufficed, my dear."

"What is it, Tina?" Samantha waved at the smoke issuing from Al's cigar with a scowl. At least he *thought* she was waving at the smoke; Goohie had stood up directly opposite her.

"Well, I think we've been banging our heads against the wall for no reason. When you guys do work on your laptops and try to process a lot of data, the system will, like, hesitate for a little while, right? When it's finished, things start working at normal speed."

Gooshie shook his head. "I know what you're getting at, but nothing could keep Ziggy locked up for this long. Her programming's too sophisticated."

"I think she may be right," Samantha said. "And the proof's been staring us in the face the whole time, but we didn't see it."

"See what?" Al was ready to do some kicking of his own, only it wouldn't be shins he aimed for. "What are you people getting at?"

"Look there." All eyes followed Samantha's raised arm to the orb suspended on the opposite wall. "That thing usually gives a better light show than a Pink Floyd concert."

Gooshie cocked his head away from the interface and looked askance at Samantha. "Pink what?"

"Floyd. They're a rock-and-roll band that—"

"*Never mind* the music lesson," Al shot. "What are—"

"Rock and roll?" Gooshie piped up, eyes alight. "Like the Dave Clark Five?"

"Well, not exactly."

"*Yes*, exactly." Al overrode Sammy-Jo's comment, fixing Gooshie with a stare. "Just like a Dave Clark Five song, that thing is usually very light and very, *very* sparkly." How out of touch could the man be?

"Well of *course* it is." Gooshie shook his head and spoke to Al in a tone usually reserved for confused five-year-olds. "But they're not just sparks. That's the main structural processing interface where Ziggy's electrical impulses are channeled through her neurochips."

"*Impulses*," Samantha cut in, "that are barely moving now. We all assumed it was a result of system damage, but I think Tina could be right."

Al was beginning to understand. What usually looked like an electric cyclone was no more than a drizzle. "Are you saying she bit off more than she can chew?"

"In a way," Tina said. "There's probably nothing wrong with the system at all. It's simply busy. *Very* busy."

"So what can we do?"

All eyes fell on Gooshie. The little man cringed slightly under the scrutiny, eyes darting from side to side. "If that *is* the case, then there's *nothing* we can do, short of rebooting her. If we attempt that, whatever she's working on will be lost. I also don't know what effect it will have on our link to Dr. Beckett. We may lose him for good."

Al bit clean through his Chivello. *Terrific!* He spit the tip into his palm. "There *has* to be something. Maybe we can feed her more information or something."

Gooshie shook his head. "Won't work unless we can find out what problem she's caught up in. And there's no way to do that unless she tells us."

"Ann-Marie Renerie." The words popped into Al's head and tumbled out of his mouth before he was even conscious of saying them. *Old age is catching up to you, Calavicci.* Al threw his cigar down in disgust and spit out bits of tobacco. "It has something to do with Ann-Marie Renerie."

"That would be my guess, too," Gooshie said.

"Wasn't that our last Visitor?"

"Yes. And before things went completely ca-ca, Ziggy kept harping on her. She was saying something about getting conflicting histories."

"That's impossible," Sammy-Jo said in the same tone her father might use, Al thought.

"That's exactly what Sam said," Al replied. "And that's why Gooshie and me didn't worry about it too much. We just figured it was witness relocation or something giving her a hard time."

"Well, my vote is for 'or something,' " Samantha said. "Al, if we *are* dealing with the onset of a conflict in time, the implications are enormous."

"More enormous than your species is capable of fully comprehending, Dr. Fuller."

Every head snapped toward the glowing ball on the other side of the room. The electrical impulses were bathing the Control Room with their chaotic patterns once again.

"Ziggy! What the hell is going on?"

"You were right, Admiral. The problem lies with Ann-Marie Renerie. She is causing a . . . how can I put this colloquially? You humans seem to like using the word 'glitch.' She is causing a glitch in the space-time continuum."

"Huh?"

"I have been occupied in codifying data I have been receiving from a dimensional divergence that has arisen since Ms. Renerie returned to her own time."

Al's sudden chill was caused by much more than the computer's cool tone. "And this divergence is being caused by . . . ?"

"The murder of Dr. Beckett."

"Time to review what we know, people." Al paced around the conference room, sweat gathering under his arms. He turned his attention to Beeks. "We need to know everything about Ann-Marie. You were the only one who had any real contact with her, Verbeena. Assessment."

"I barred access to the Waiting Room with good reason." Verbeena passed around folders containing copies of Ann-Marie's psychological profile. "The second she got here, she was extremely volatile, as Dr. Fuller can attest. We had to sedate her right away. I'd say that she was one of the most unstable Visitors we've ever had, including Leon Styles. The only difference between the two was that she didn't have a gun."

"She couldn't have been *that* bad," Al said, absently patting his ribs. They still twinged whenever he took a deep breath.

"In certain ways, she was worse," Verbeena replied. "Styles's violent behavior could be attributed to a history of neglect compounded by a chemical imbalance. Ann-Marie, on the other hand, had no solid basis for her behavior that I was able to see. In tabloid terms, she was a classic case of the good girl gone bad. Based on what Ziggy could dig up, all factors pointed to an upper-middle-class upbringing with a stable family life. She even has a degree in art history. That's what makes cases like hers so enigmatic."

"Let's cut to the chase, Verbeena," Al said more testily than he wanted to. "What could drive her to murder?"

"There are no deep psychological mysteries there. She was a very greedy, self-centered individual. And by 'self-centered,' I don't mean conceited. She shaped her whole outlook on the basis that she was the most important thing in the universe. Everything else came second. That's why seeing Sam's reflection in the mirror was so traumatic for her. It threw her belief system a curve." Verbeena sighed. "And I'd bet that's what created the problem we're facing now. We've planted a time bomb, and there's no telling how or when it could go off."

"That is incorrect, Dr. Beeks," Ziggy replied serenely. "As I said before, events are unfolding that lead to the murder of Dr. Beckett."

Gooshie looked up from his folder. "Are you implying, Dr. Beeks, that she has gained a foreknowledge of the Project and is using murder to counteract the changes Dr. Beckett made in her life?"

Verbeena shook her head. "I think that's giving her too much credit. She doesn't know much about the Project, and I never told her about the time-travel aspect. No, I'd say the most likely scenario is that she has regained her Leap memory in some way. It's foreknowledge of a sort, I suppose, but I doubt she thinks of it in those terms. She's more likely driven by revenge."

"Okay, so we've established motive," Al said. "But what I don't understand is, if she *does* murder Sam, won't it have already taken place from our perspective? Wouldn't the change have been instantaneous the second she Leaped back?"

"It's not that simple Al," Samantha said. "That's been bothering me as well, but we're dealing with a lot of variables here, all of which must be decided in order for a new time line to take root. The easiest example I can think of is Dr. Beckett's Leaps. If changes in the continuum *were* instantaneous, then we would know the ultimate outcome of his actions the very second he arrived in a particular time and place. Instead, we deal with percentages based on countless variances in probability. Those variances are the reason we're all still here."

Al's head was swimming in the attempt to digest it all. "So we're coasting along on a slice of probability that's shrinking with each passing minute." Terror snaked its way up Al's spine like a choking vine. His primary concern was saving his best friend—and Sammy-Jo Fuller, for that matter. If Sam was killed, nature would snip her out of the pattern like a botched seam.

But try as he might, he couldn't suppress the other cowardly thought whispering to him from the recesses of his mind. He knew all too well how his life would have turned out without Sam. He had already trod far down the road of alcoholism before Sam provided a detour. He would have been dead by this point without the Project. Or, worse, a pathetic old drunk that everyone pitied. "We need to act, people. I'm open to suggestions."

"The first thing we need to do is find the time and place

of the murder," Samantha said. "How 'bout it Ziggy?"

"The fluctuation of the time stream has prevented me from ascertaining that information, Dr. Fuller."

"How long will it take you find out?"

"Unknown."

"Ziggy, I don't want to hear another word from you until you can tell us where to focus our efforts." There was no hint of Al in that tone; Admiral Calavicci had taken over. "Take all but the most vital systems off-line if you need to. If I don't get an answer by sunrise, you won't have to wait for an alternate time line to do you in. I'll find the socket you're connected to and unplug you myself."

The lights in the conference room dimmed immediately. Al nodded with satisfaction and addressed the others. "Once we find out where and when, what's our next step?"

"Two possibilities," Samantha said. "First, we figure out if any of Dr. Beckett's previous Leaps correspond with that point in time. If so, we may be able to find some way to modify the Imaging Chamber's programming to lock onto him at said point and have him stop the event himself."

"Gooshie, can you do that?"

"I may be able to find some theoretical foundation to support the idea," Gooshie answered. "But I don't think we'll have much luck."

"I don't care," Al said. "Get to work on it. Tina, you too. If the Imaging Chamber needs to be modified to get it done, tear it down and start from the bottom up if you have to. Keep me posted. Dismissed."

Gooshie and Tina practically ran from the room, throwing ideas back and forth before they reached the door.

"What's our other option?" Al asked Samantha.

"One of us Leaps and stops her."

Al considered. "Before we go any further on that idea, we need to get committee approval." He loathed the thought of trying to explain all this to Weitzman, but it was a necessary evil. He might have a chance with McBride. She was pretty sharp.

It struck Al that she was another one whose life would change if Sam were killed. The negative ripples kept spreading further from the center. There was no way of telling when they would stop, if ever. "If I'm gonna face them, I need to shower and get into uniform. Set up a satellite link and patch it through to my quarters as soon as you can."

Samantha Fuller was out the door, with Al on her heels, before Beeks cleared her throat loudly enough to stop him. "Aren't you forgetting one thing, Admiral?"

"What?"

"*Sam*. He hasn't had any contact in days."

Al slapped his forehead and grimaced. "Geez! What a *stunad*. How could I forget? Still, he'll have to wait just a bit longer. The committee will be on-line in too short a time for me to do him much good. Look, dig some odds out of Ziggy and prepare a brief for me. Sam was saying something about a town council meeting when I last saw him. He may have the problem solved by now. But if he doesn't, I want to go to him with every possible option."

An instant later he was in his quarters, out of his clothes, and taking the Naviest of Navy showers. He might not mind McBride catching him in a towel, but Weitzman was a different story.

As it was, he only had time to get into the top half of his uniform before his computer blipped at him. He sat in his skivvies, feeling like a dumb anchorman joke as he accessed the link.

The screen was divided into four smaller screens. He was going to give Sammy-Jo a big kiss for contacting only those members absolutely necessary to make this decision. Maybe this would be easier than he expected.

Then again, maybe not. A closer look at Senators Weitzman, McBride, Turner, and Diliberto revealed a bleary-eyed and none-too-happy bunch.

"This better be good, Calavicci," Weitzman said, absently rubbing at a bandage on his cheek. So the rumors that he had had a mole surgically implanted *were* true. Al fought temptation and got down to business.

"It's actually very bad, Senator. We're facing a problem the likes of which we've never seen, and unless we act immediately, the consequences will be dire." Al gave them the basics.

"That's one fantastic story, Admiral Calavicci," McBride

said. She was the only one who didn't look mystified as the details were given. "But I, for one, feel that Leaping another person is out of the question. There are too many risks."

"Senator McBride is right," Turner said, smoothing his white goatee into a point. It was just his luck, Al thought, to be strapped to senators with egos bigger than most rock stars'. But at least Weitzman had chosen a former president to model himself after; Turner could have been Colonel Sanders' evil twin. "You're basing your conclusions, sir," his drawl elongated the word to *saaah*, "on information that is coming from a dubious source at best."

"I would hardly call the world's most advanced computer system dubious, Senator," Al replied. *Especially not where she can hear you, Foghorn.* "But the fact remains, something must be done. And if it comes down to sending someone back to stop it, the risk will be on my shoulders. I'm the only one who could do it."

"Your dedication is commendable, Admiral," Diliberto said. No bolo or top hat here; Diliberto was perhaps the sharpest and most laid-back senator Al had ever met. Ego didn't enter into his equations, only results did. Of course, that in itself was a two-edged sword. "But I think there's one possibility that you're not voicing, and I doubt you've overlooked it. Frankly, the risks you would be taking are a secondary concern in this case. The real question lies with the person you displace. Project Quantum Leap has already turned from an experiment of observation to one that's invasive to private citizens, good deeds or not. I, for one, can't sanction tearing more people out of their lives so you can clean up your mistakes. I think I speak for all of us when I say that."

"I understand your concern, Senator," Al said, feeling the sting of that second edge, "but the Leap would be a one-time event targeted to a specific individ—"

"You can guarantee that, Admiral?" Turner accentuated his words with the crook of his cane. "Time to face facts. If you can't retrieve Dr. Beckett, what will make your Leap any different? You might be stranded in time just as he is. Then who would run the Project?"

They were worried about who would run the *Project*?

Were they blind? "Senators, I'm confident that we can work out all of those details on this end. We can't let them distract us from the real issue. Dr. Beckett's *life* is at stake, and we have to do everything in our power to prevent his death."

"It's this committee that decides the extent of your power, Admiral," Weitzman said. "Just remember that. And we all acknowledge that it would be a great tragedy to lose such a brilliant man as Dr. Beckett. Do anything you can to save him. But heed my words well; under no circumstances are you to Leap yourself. Let's make it official, folks. Senator Diliberto?"

"Agreed."

"Senator Turner?"

"Agreed."

"Senator Laugen?"

"Agreed."

Laugen? Al focused on the corner of the screen where McBride had been. A blonde man stared back at him. Where the hell had this big-headed Ken doll come from?

It's starting already. How much time do we have left?

Al couldn't waste any more of it on these nozzles. They were bringing a literal meaning to the phrase "death by committee." "I understand, Senators. Calavicci out."

He severed the connection and took a deep breath. To hell with etiquette. What else had already changed? Reality could be flickering back and forth like a bad lightbulb without any of them even realizing it. How long before the new reality stayed put?

He jumped into his pants and ran to the Control Room.

"What's the prognosis, Gooshie?"

"Not good, Admiral. It could be weeks before we even come up with a test program, and weeks more before we work the bugs out. The numbers just won't crunch."

"Forget about it, then." There was no time left. To hell with the committee. Court-martial was a small price to pay. "Juice up the Accelerator Chamber. Tina, get me a Fermi suit."

CHAPTER TWELVE

"To hell with him. It's the price I named or the show is off. I don't think I'll have a problem finding another buyer."

Rodimer studied his shoes. "He said something about calling the cops."

Ann-Marie walked around to the front of the desk. "Is that so?" Her tone made the courier lift his head. "Well, tell Simmons that if he wants to screw with me, I'm sure the cops will be real interested in his private collection. Always repay threats with bigger threats. That's your lesson for today. Now go tell him. You'll see how his tune changes. If I don't hear back from him in forty-five minutes, his package goes to the next in line." She walked the man out of her office and across the shop. She unlocked the door. "And if you use a phone within fifteen blocks of here, I never saw you before."

"I know the drill," Rodimer said testily, "and I'm getting sick of it. When do I get to make a deal on my own?"

"When I think you can handle it, Billy Boy, that's when.

If you don't like it, I'm sure the McDonalds on West Fourth is hiring. Out.''

Rodimer sulked out onto the sidewalk and Ann-Marie bolted the door behind him. "Damn kid is too eager for his own good. And how about that Simmons, threatening to call the police? Well, if it's about breaking balls these days, I can be the biggest ball breaker of them all."

Fight, child, but keep the greater prize in mind.

"The greater prize," Ann-Marie mumbled to no one.

That prize, that face, was the *only* thing that kept her fighting at times. She had struggled more in the last two years than she had in her entire life. But the gods were on her side—of that, she was sure. She had found out as much about them and the religion Evangelene had introduced her to as she could after she got out. It wasn't an easy task.

It was called Santeria. Some sources said it was basically Caribbean, others said it had Latin American origins. It was linked to voodoo, strangely tied to Catholicism, and was a way of life for a lot of Cuban natives. She could find only one book that gave it African roots, saying the slave trade introduced it around the world.

Only the *Orishas*, the powers that ruled over the mania of nature and humanity, remained constant. They were what mattered to Ann-Marie. To harness their favor was to harness the world. They had shown her favor in many ways.

For one, the feds had missed a few of her bank accounts. The money didn't amount to much more than a sneeze, really, even with a decade's worth of interest tacked on, but it was enough to establish the shop and conduct legitimate business while she got her feet under her again.

She walked through the shop, past relics of a slightly older time: a German headboard here, an Italian dresser there, some early American pieces. Not great stuff, even for a small-time Greenwich Village antiques dealer, but she felt some affinity for it. And it was enough to fool the world.

The gods had made that easier as well, seeing to it that she served her full sentence. No parole officer came to check on her, no restrictions were placed on her movements. She was rehabilitated in the eyes of the law, no longer a threat to society. Nice and invisible.

She planned on staying that way. The business had changed a lot while she was on the inside, had become even more ruthless. No one worked in set circles any longer. Smugglers were taking their cues from the Fortune 500 these days; the only code left was hooray-for-me-and-the-hell-with-you. It had taken her this long to reestablish old contacts and make new ones that she could trust, if she could even call it that. Drops were the only thing she handled personally, and not even those were face to face.

Rodimer made that easier. He was fool enough to do anything she said. He picked up the "special" parcels delivered to the empty apartment in Jersey, made all the phone calls, and was hungry enough to hang on with no more than an empty promise or two. The perfect patsy if things got hot.

She reached her office and gave a last glance to the quiet street through the front window before closing the door and retrieving Simmons' package from the corner.

The crate's lightness belied the treasure within. All the transactions she had made since getting out were dry runs to prepare for this. It was her first important deal, her first *real* move toward regaining all that she had lost—her first step in destroying him.

You rise, he falls. It's nature reclaimin' her proper order.

"He falls," she repeated as Evangelene's voice came again. At least she *thought* it was her friend's voice. It was darker, more sinister, but it held the same Southern cadence. Who else could it be?

Ann-Marie dug the claw of the hammer behind the wood and pulled. The nails made a satisfying squeal, and the top of the crate popped away. She threw it to the floor and dug through the excelsior.

The small footstool inside was seventeenth-century French. She already had a potential buyer for the piece, which would fetch five hundred, easy. A small bonus.

She cleared the desk and flipped the stool over, prying the legs carefully from the base and setting the frame aside. The tacks keeping the heavy floral fabric in place pulled out easily. Her heart raced, but she forced herself to work carefully.

The cambric gave way to a pine base. She reached underneath, between the ancient horsehair and cotton batting, and slid out a plastic pouch. She put the package aside and reassembled the stool.

The shop's legitimacy had just become a charade, and maintaining it was more important than ever. She filled out a tag, attached it to the stool, and found a spot for it out front. Business attended to, she returned to the office and took the package off her desk, unbolted the door behind the desk, and descended into the safety of her room.

The smell of hot wax intruded on the musty dampness, flickering flames casting their soft glow around the cellar. The candles, lit twenty-four hours a day, were grouped in the colors pleasing to the different *Orishas*—red, black, and white for Elegba; green and black for Ogún; blue and white for Yemayá; pure white for Obatalá—currying their constant favor, their all-seeing power.

Roaches scurried underfoot as Ann-Marie made her way across the cracked concrete floor to the night table beside her bunk and switched on the lamp there. Sketches of Mirror Man overlapped on the walls, their variety ever-increasing, ever-reminding. She had spent endless nights scratching them out, preventing herself from forgetting him.

The converted storage room was a replica of her cell, but it was a prison of her own design and it served a purpose. That made all the difference. Its dankness fostered the nightmares, kept the pain fresh, alive.

It kept her focused on the greater purpose.

You remain weak so long as he lives, and weaklings are never free.

She cringed in the gloomy space. It was the other voice, the one she didn't like—the one that always chided, always pointed out how worthless and ineffectual she had become. She tried not to listen to it, squeezed her ears against it so hard at times that she thought they would bleed. It never worked.

She sat on the bunk and unsealed the pouch, drawing out a cardboard sheath.

Inside, sandwiched between pieces of acid-free paper, were four of the most beautiful works of art she had ever handled. Each ancient piece of parchment, heavy and yet fragile, depicted a scene from the New Testament, colorful illustrations by a monk long dead. The book from which they had been cut was at least a millennium old, collecting dust in the holdings of a cathedral library in Spain.

The art historian within her took in their splendor, grieved for the mutilation of a priceless artifact. The other factions of her mind, writhing in hate and shame, saw new possibilities. Five hundred thousand of them, to be exact.

Ann-Marie laughed, a faint reflection of the sounds careening through her head. With this deal, she would finally have the resources to hire someone, maybe several someones, to find Mirror Man, to attach a name to the face that had dangled like a carrot tied to a stick in her mind's eye these many years. His features had fought her since she had left prison, had tried to take on a ghostly quality. But she fought back, holding tight to every detail, every line, every hair. Even as she did, he lingered in the corridors of her mind, laughing at her struggle, at its futility.

How could she possibly find him? Even if she had an empire's worth of money at her disposal, the only lead she had to go on were fevered night drawings. She even found herself wondering if he was real at times. Had it all been a delusion?

No! No doubts. She sprang from the cot, bearing down on the nearest sketch. He *had* to be real. She needed him to be real, needed to see him destroyed. Otherwise, the voices would never leave her alone.

"It won't be long now, Dr. Frankenstein," she said softly, caressing the image. "How will your screams sound, I wonder?" She looked at the face for a long time, absorbing every detail in her mind, before she retrieved the manuscript pages and left the room.

The parchment rustling in her hands bolstered her spirits. They were victory enough in themselves. But she forced self-control. She would permit herself to celebrate only when his death-glazed eyes could bear witness.

Ann-Marie took a cardboard tube from her desk and carefully rolled the yellowed parchments, depositing them inside. The phone rang a few minutes later. Simmons already? She swore Rodimer could fly sometimes. She picked up the receiver. "Antique Village."

"It's a go." The phone clicked down on the other end.

She deposited the tube into her purse and put on her coat and hat, then checked her pockets for tokens before pulling on her gloves. A short hop on the subway was all that remained. She walked up Perry Street, admiring the elegant brownstone apartments draped in ivy and wrought-iron, the mystical castles of New York's quieter royalty.

She thought of her main clientele, the assholes who preened on the Upper West Side, draping their fortunes from their park-view terraces and hiding behind legions of doormen—the faux-rich and the Euro-trash, all having about as much real class as a Queens lottery winner. But not her. The Village was a model of subdued riches, set amid drag queens and discount record stores, dark cafés and tacky T-shirt shops. It was alive, spicy, what she ached to be. Immersing herself in it was another reminder, another way never to forget how dead and empty he had made it all.

She turned down Christopher Street and stalked to the subway station, then descended to the platform. The urine-scented air and filthy gray concrete underfoot reminded her of prison. The train's sliding roar brought to mind automated barred doors rolling closed. Even the soft dinging indicating the closing of the train's doors seemed an homage to prison's ever-present bells. The ride uptown was peaceful, however; Evangelene never joined her in the subway.

Traveling through one dark tunnel after another, she finally reached her destination. The Port Authority terminal was damp and crowded, milling with uniformed workers, homeless wastes and the poor fools of the world who had no choice but to travel by bus.

She walked leisurely along the broad corridors toward the banks of lockers, even though she had taken a local train that stopped at every station. She wanted to be sure Rodimer had enough time to do his job.

When she finally reached her destination, she looked for locker 64. She had chosen it because the number was etched on her brain, the first part of her prison identity. No one else could use it because Rodimer always held on to the key. She groped in her purse, finding the duplicate she had made, and slid it home with a twist of its orange head.

An attaché case was inside. She drew it out and put the tube in the locker, then closed the locker and deposited enough coins in the slot to get her key back. Rodimer would return for the tube shortly and deliver it to Simmons. Her risk factor stayed minimal, and if Billy Boy got pinched, her key would go sailing off the 59th Street Bridge.

He says I had a duplicate key made, Officer? Isn't that sort of thing illegal? I only hired him to run errands a few days a week. . . .

She laughed into the surging mass of people, enjoying the thought. Maybe it wouldn't be *that* easy, but there wasn't a shred of physical evidence that could link her to the stolen pages unless someone was trailing her right this second with a hidden camera. Laundering the cash came next. As easy as Rosa with a CO.

Years of buildup ended in the weight dangling from her fingers. She was no longer on the outside looking in, or on the inside longing to get out. She was a true player, something she had not been even in the days before prison, before her conflict with the good doctor. But it wasn't enough, wouldn't be enough until she saw fear in his eyes again. *His* terror, this time.

Day by day we're gettin' closer. We'll be a-singin' in the clouds while he's a-dyin' in the dirt.

"You said it!"

Hope is a disease of fools. Fool.

"No! His time will come! Evangelene says so! The *Orishas* will make it so!"

Ann-Marie was oblivious to the stares she drew, the voices spurring her on past the subway entrance. She would take a cab back to the shop. It would mean a witness who could put her at this place at this time, but you didn't board a subway train carrying a half-million dollars unless you were insane.

She joined the taxi line on the sidewalk, lifting her collar and pulling down her hat. She studied the sidewalk, a nondescript mass of winter wear turned against the chilly October afternoon.

The next cab was hers. As it pulled up, she caught her fish-eyed reflection, and another, in its window. She went cold, felt the terror of the blue room. Mirror Man's face hovered beside hers, washed ghostly in the glass. She fingered the wraith, the world going mute, then wheeled in an explosion of ire. Her free hand reached up in an involuntary claw.

Kill!

A glossy picture of the Mirror Man confronted her, his smile framed by a red border and the fingers of the man holding him up. *Time* magazine. It was a copy of *Time* being read by the man waiting behind her.

"Man of the Year?"

The cabdriver beeped the horn with impatience, and the man peered at her over the edge of the open pages, eyebrows raised in a question. "Can I help you?"

"You just did." She grabbed the periodical and hopped into the cab. "Perry and Bank," she said absently, and the driver pulled into the honking stream.

She took in his image, stunned beyond words. Part of her mind howled. Another part wept in joy. The only sounds she could seem to manage were giggle after stupid giggle.

"Sam Beckett," she finally managed. The unremarkable words rolled off her tongue, awkward. "Sam Beckett," she repeated. Not very exotic, not as dangerous as she imagined it would be. It was a name as plain and real as his nondescript features, as stupid as the white lock of hair falling across his forehead.

Droplets fell on the page, distorting his smile. The adrenaline coursing through her made Ann-Marie shaky, nauseated, and she rolled down the window.

"Please, the heat."

"Huh?" she stared toward the front seat.

The driver studied her in the rearview mirror, eyes curious. "You're letting the heat out. It's cold."

"I was just having a smoke," she replied dumbly, fishing the cigarettes out of her purse and bringing one to her lips. "Want one?"

The driver shook his head and drove on in silence.

Ann-Marie kept her eyes focused on the passing traffic. The prize sitting in her lap terrified her. She didn't know why.

Coward.

"It's not supposed to be like this," she mumbled. All the dreams she had had of this moment never included such queasy dread. *"He's* the one who needs to be scared." She took a nervous drag on the cigarette, hand trembling.

What's the problem? Still bowin' to fear, child? It's easy to swear revenge against a fool fantasy, to cuss at pictures in the dark. You gonna let a little flesh an' blood stop you now?

"No!" Ann-Marie pitched the cigarette. One more puff would make her gag. She swept the magazine out of her lap, emotions seething. "I'm not afraid of you."

"Are you okay, lady?"

Ann-Marie looked up, saw the driver staring at her once more as he brought the cab to a halt. Washington Square Park lay directly in front of them, the low afternoon sun casting the arch in glaring relief. "Here. I'll get out here." She found a ten in her purse and shoved it through the slot in the cab's protective divider before stepping out, magazine clutched to her chest.

She walked through the park and read. A farm boy, a prodigy, a musician, a doctor in seven different ways, five of which had to do with physics (none in sociology), the next Albert Einstein. Ann-Marie's mood worsened as she learned more. A regular Mr. All-Fucking-American, this one. And not only was he Man of the Year, he was next in line for the Nobel Prize for his *revolutionary* work in physics.

She reread the article twice on the way back to the shop, each go-round making her madder and madder because she could find no mention of the work connecting him to her. The closest thing was a line saying he had been associated with "various government projects." Had no one exposed his heinous crime?

There was also an interview with one of his college professors. She skimmed the piece, looking for the details she needed, but soon closed the magazine in disgust. *Another* one who couldn't gush enough, it seemed.

Ann-Marie walked on in a haze of frustration and bewilderment. How could a monster warrant such accolades?

When she finally turned the key in the shop's front door,

the magazine was little more than a ball of crumpled pages in her fist. She stalked back to her office, turned on the light, and threw the copy of *Time* at her desk. It careened off the blotter and hit the wall, pages spreading sloppily on the floor.

She was on her way to pick it up when the phone rang.

"Hello?" she barked.

"Did you get it?"

"Oh, I got him all right." How did Rodimer know about him? She always kept her room locked. "How did you find out?" If he had been snooping around. . . .

"What do you *mean*? I'm the one who dropped it off."

The phone clunked noisily on the wooden desk and Ann-Marie sat down hard. It replayed in her mind slowly, painfully, excruciating in its detail.

Searching through her purse for cab fare. Her eyes moving back and forth, sweeping across the briefcase on the floor next to her feet. Sweeping, but not seeing. Climbing out of the cab with only the magazine.

He had done it to her again.

The voices screamed. She screamed with them, holding her ears against the terrible noise within and without. A half-million dollars, touring the city in a smelly cab until it found a new owner. She would never see it again. Hell, she hadn't even seen it a first time!

Her fear, her confusion, her anger, her pain hardened into a cold ball, pulsating in her middle with wave upon wave of hate. It brought with it a strange calm, lent a new determination. She retrieved the magazine from the floor behind her and smoothed the pages, looking for a certain article as she picked up the receiver once again.

". . . the hell is going on?"

"Shut up and listen. When you come back, have a gun with you, or don't bother coming back at all. Ammo, too." She dropped the receiver, breaking the connection.

As much as the money was worth, she told herself again and again, it was really meaningless. She had desired it only as a means to find him. That end had been achieved, and it took only one deal to do it. The gods were indeed showing her favor.

She riffled through the magazine in front of her. Where

was it? Pages tore under her hands. There. The professor's interview. He was planning a celebration, a small shindig for Dr. Beckett. She closed the magazine and stared at him, drank him in.

"Doctor Beckett." She enjoyed the foulness of the words on her tongue. The rock in her stomach convulsed faster, provided strength. "My dear Sam. How will your screams sound, I wonder?"

CHAPTER THIRTEEN

"We all know dirt and politics go together, but *this* Arkansas candidate brings new meaning to the term 'mud-slinging'—"

Click.

"*She's* a mud queen; *he's* squeaky clean. But now *both* are getting down and dirty in this small town showdown—"

Click.

"Politics is a slick game, but *this* mud-wrestling mayoral hopeful knows all about slippery conditions—"

Click.

Sam turned off the television and sipped his ginger ale in shock. "I can't believe what's going on here."

"Why did you turn it off?" Kat asked, reaching for the remote. "I want to see how they put it all together."

"Don't worry." Sam held the clicker out of reach. "The other girls are all on VCR duty. You'll be able to show the tapes to your grandchildren. We need to practice some more."

"I don't *have* grandchildren. And I'm not likely to, either, unless God is in the mood for a miracle. Now hand over that remote or get out of my bar—I mean campaign headquarters."

Sam shook his head. "No can do. You said yourself, things are moving quickly. Now that Barrenger wants a debate, you have to be ready. He'll beat you otherwise. Then those stories will be the *only* thing you have."

"We've been over it and over it," Kat said, getting up.

"And we'll go over it again." Sam tried to keep his tone neutral. "Press doesn't mean votes, and the town won't put you in office unless you can show people you're about more than keeping this bar open."

"Right now, you should be satisfied that I'm keeping my hands from around your throat," Kat shot back, flipping a cigarette from the pack lying on the bar. "Doing this was *never* about keeping the club open; you know that. There's lots of places where I can take my business and be welcomed with open arms."

"It's not a question of whether or not we know it." Agitation was creeping into Sam's tone. "You have to make the town know it, too. And you have to do it with Barrenger trying to trip you up the entire time. He can turn your words around and make a fool out of you without even breaking a sweat." He snatched the lighter out of her reach. "Please don't light that here."

Kat tapped the cigarette on the bar and stood. "I'm taking five."

"Make it three." Sam smiled. "I'll be waiting."

She retreated to the office, scowling as she lit up. No Smoking was a rule he had imposed early on. If he had to prep the woman for this election, he didn't see why he had to choke doing it. Sure, it interrupted their prep time and made Kat irritable, but at least he could breathe. They had little enough time for breathing.

Things were, indeed, moving quickly; by the afternoon following Sam's initial suggestion, the girls had canvassed the town and collected enough signatures on petitions to put Kat officially on the ballot.

The decision to set up a debate was also impromptu; once Barrenger had heard of Kat's candidacy, he insisted on it.

Apparently the councilman hadn't been satisfied with the coverage of the council meeting (not one station had aired his Churchill quotation) and was looking for a forum where he could pit himself against Kat's evil influence, most likely in a frenzy of exaggerated gestures and blustery sound bites.

Things *really* began to happen, though, when one of the networks picked the story up from a local affiliate and ran it to end the national newscast. The afternoon talk shows then got hold of it. Within two days Wilson was swamped with news crews from around the country, a town shot headfirst into the maw of tabloid television.

The shows he had just deprived Kat of seemed primitive to Sam; by the time he had first Leaped, news in the 1990s was being driven by slick, first-run syndicated "infotainment" shows like *Hard Copy* and *Inside Edition*. But the magazine shows were still floundering in the late 1980s, still taking time to find their niche before their incredible proliferation took place.

Of course, the content hadn't changed much; just viewer's standards. All of the programming was driven by sensationalism, dishing out a steady diet of juicy celebrity tidbits, the occasional royal scandal, and stories about regular people caught in situations offbeat enough to be compelling.

Kat's story fit the bill perfectly. It had politics. It had intrigue. It had conflict. And, most important, it had girls in bikinis.

The opportunities weren't lost on Kat. She took to the coverage like a preacher to a congregation, playing her knowledge of exploitation techniques like a finely tuned instrument. She insisted that the girls wear bathing suits and accompany her to the barrage of interviews she was doing. The camera's eye had found her, and she wanted to keep its attention as long as she could. At least she was intent on winning.

The roller-coaster ride they were on was exciting, far more than he ever anticipated; but Sam, having spent the better part of the last two days in a bikini and high heels, was getting a bit queasy and irritated. Kat was paying too much attention to the sizzle and not enough to the steak. They had drilled for hours, but she was still barely prepared for her confrontation with Barrenger in three nights' time. If she really wanted a shot at winning, she would have to convince the townspeople she could be a good mayor. They were the ones who held the chips, and he knew that if the whole country was watching the power clashes of this town, then its citizens should turn out in full force at the polls.

Support was high, for now. It all hinged on the debate. It would either keep the momentum rolling or stop it dead. That realization was what had Kat on edge. Despite the enthusiasm she displayed for the media, he knew she was terrified of confronting the councilman in a debate. It was like pitting a street brawler against a black belt; the thug might win, but the odds were long.

Sam's stake in shortening them was biggest of all.

That was the root of his worsening mood. The specter of spending the rest of his life as Candy Apple was continually looming at the edges of his thoughts. Kat's seeming indifference to the issues made the possibility all the more likely.

There was only one thing that could dispel it. But if he saw Al right now, Sam would be more angry than relieved. No explanation he offered would be good enough. A childish reaction, maybe. But hologram or not, Sam was going to kick the Observer where the sun didn't shine.

"Let's get started again." Kat resumed her seat, reeking of smoke and perfume. The combination didn't improve Sam's mood. "And there's no reason to look like you're going to kill me. I know you're right about the debate. It's just that it's been a while since the public showed any interest in me, and I've *never* been this mainstream. I forgot how much like a drug it could be. Permit an old has-been her fix."

"You may be many things," Sam said, mood lightening, "but you're no has-been. Has-beens cling to past glories instead of focusing on new challenges. I don't think you have that problem."

"Such as the new challenges Wilson will face in the shrinking economy of the next decade?"

"*Now* you're talking," Sam said. "Keep the strategy in mind—focus on the issues. That way, if he attacks your character, it'll look like he's avoiding what's important." A questionable strategy in politics, taking the high road. But it was a road he was familiar with, and Sam saw it as the best option. "Let's hear some predictions and solutions. Remember, be—Gooshie!"

The man's form blinked into existence a few feet behind Kat. Gooshie stopped his study of the hand link upon hearing his name and looked around as if he were just registering his surroundings. "Dr. Beckett! I've achieved a lock, Ziggy. Keep this hologram as stable as you can." He ran his fingers through the bar. "Remarkable...."

"I think a reserved attitude might be better, don't you?"

Sam snapped his eyes back to Kat. "Huh? Oh, yeah ... yeah. You're right, reserved is better. I have to use the bathroom. Excuse me." Sam practically ran from the room.

"You're only human, honey," Kat yelled after him. "Next time, don't hold it so long."

Gooshie flashed in next to him, but Sam kept moving, along the short hallway and into the lounge. He wanted as many walls between him and Kat as possible. He could feel some yelling coming on. He'd have taken the conversation out into the chilly evening if he were wearing more than a string bikini.

"Where the hell have you been?"

"It's a bit of a long story," Gooshie chortled, quickly looking down at the hand link. "Ziggy's...." More laughter. "Ziggy's had some problems and...." His words dissolved into a stream of guffaws so strong that he dropped the hand link while clutching his sides.

Sam waited. When Gooshie finally popped back into view, he was wiping tears from his eyes and breathing heavily. "Sorry, Dr. Beckett," he managed. "Sorry."

"What the hell is so funny?"

"Your clothing is rather ... becoming. Not the habiliments of a quantum physicist, perhaps, but yellow is definitely your color."

"Okay, okay. That's enough." Sam could feel himself flushing, and it had little to do with anger. He could only *imagine* what Al would have said. "What was it you were saying about Ziggy?"

"We had a glitch in the system."

"What's the problem? Where's Al?"

"He was . . . called away unexpectedly." Gooshie's image blitzed to one side, then snapped back into place. "And the problem is being taken care of. How are things progressing on this end?"

"It's customary for *you* to tell *me*."

"Forgive me, Doctor, for being at a slight disadvantage. We've been running the Project at minimal capacity for a few days. With everything that's happening, it's lucky that we've been able to contact you at all. You need to bring us up to speed before we can do the same for you."

Sam outlined the events that had taken place since he last saw Al. Most of his tension drained away as he did so, but an ominous feeling lingered. What kind of glitch could halt contact for so long? Where did Al have to go in such a hurry that he couldn't warn Sam beforehand? He asked Gooshie as much when he was finished.

"It had to do with one of his ex-wives, I think," Gooshie said, plugging the information into the hand link. "And as for Ziggy, you'd have to ask Tina exactly what the problem was. Something about the hardware, I believe. But it's all fixed now."

"Wait a minute. First you say the problem is still being taken care of, then you say it's fixed. Which is it? What's going on back there?"

"I, uh . . . just meant that the problem has been fixed to a point where we can reach you again. As I said, it has to do with the hardware. . . ."

"And nothing else?"

"No."

Sam was having trouble buying it. The frustrating part was that he didn't know why. The explanation fit, but a sixth sense told him that . . . things . . . were on the brink of change. He couldn't say *what* things exactly, but it wasn't a good change; it felt more like he was dangling by one hand, the cliff's edge crumbling in his grip. No computer problem could generate those kinds of feelings, *unless*. . . .

Catastrophic system failure? Were Ziggy's neurocomponents breaking down? That could explain the feeling of dread. There was no telling how something like that would affect him. It could strand him permanently in the aura of his host (not a pleasant prospect on this Leap); it could just as easily kill him. Without the technological tether linking him to his own time, God or Fate or Whatever might decree his presence in the past as too anomalous, too messy to fit into the pattern.

One thing was for sure—they couldn't repair the system without access to fresh samples of his cells. He would be dead, as far as the Project was concerned.

A disturbing image flashed in his mind. Al on an operating table, part of his skull removed, surgeons frantically gathering the only other cells familiar to Ziggy's matrix, the only other cells with a remote possibility of performing the tasks Sam's had, before the whole system went kaput.

Sam took a deep breath and banished the mental picture. There was no way things could have gone that far south. Gooshie would definitely tell him if the problem was that drastic. Wouldn't he?

Of course! Stop being so paranoid.

The way Gooshie seemed to be hedging, the problem was with the software. And Sam would have a better chance of giving birth (*again*) than getting him to admit it. The programmer prided himself on keeping the system bugfree, if Sam could trust his Swiss-cheesed recollection.

Gooshie faded in and out like a giant white firefly, studiously entering data into the hand link with a "Who, me?" expression on his face. That had to be the answer.

Then why did doom still feel so palpable?

"Getting something," Gooshie said. "Here we are. Well, Ziggy says the move for the mayor's office was the right one. It has the highest odds for success. However...."

"However *what*?"

"Ziggy says...." Gooshie's voice fell a few octaves and separated into four distinct harmonic patterns before becoming incomprehensible. His image faded completely, and for a long moment Sam's paranoia didn't feel ridiculous.

"... the election." The voice came from behind Sam, startling him. He turned to face Gooshie, who grinned at him.

"Dammit, Gooshie! Stop doing that! What did you say?"

"I *said* that even though Ziggy is quoting the odds in your favor, records show that Kat will still lose the election. It's odd, because all of the exit polls pointed to her as a shoo-in."

"How can that be?"

"The most likely explanation is that Barrenger has somehow rigged the election."

Sam felt the sting of three days wasted. All the prepping in the world wasn't going to help Kat if Ziggy was right. He sat down hard on the nearest couch. "So no matter what I do to get her ready, she'll still lose? What should I do?"

Gooshie moved next to him, intent on the hand link. He moved to sit down as well—and fell through the couch, landing hard on the Imaging Chamber's floor. Only the top of his head remained visible above the cushions, his eyes darting back and forth in confusion.

It was Sam's turn to laugh. "Even Al never made that rookie mistake!"

Gooshie pushed himself up, rubbing the small of his back. "A bit tricky, getting used to this whole sensation," he said defensively. "It's hard to remember that you're not real when I'm standing here talking to you."

Sam shook his head. Even if Ziggy made the hologram blink like a neon sign, Gooshie would probably still forget it wasn't real. As soon as he focused on one thing, he became oblivious to everything around him. That's what made him such a good programmer and such a lousy social being. Al always went on about how out of touch Gooshie was with reality. Sam knew he was actually *very* in touch with reality, almost ruthlessly so. He just took it one sliver at a time.

"There has to be some way around Barrenger," he said. "Come on, what's the best way to beat this thing?"

"Your only chance is to expose him as corrupt."

"How?"

"Ziggy doesn't know."

"But the election is only four days from now. I don't even know where to begin! Does she have any suggestions?"

Gooshie squinted at the tiny readout. "Get...." He whacked the hand link. Some traits carried over from Observer to Observer, apparently. "Get...."

"Get what? Evidence that he's fixing it? An informer?"

"... busy. Get busy."

"*That* helps." Sam paced in frustration. "Any clues as to how?"

"She says to find evidence that he's fixing it, or an informer," Gooshie said.

"I think it's time for you to leave, Gooshie," Sam said as calmly as he could, through gritted teeth.

Gooshie looked at him questioningly. "I believe there's another matter of concern we need to discuss. Theresa Singer and her brother. You wanted the odds on reconciliation. Ziggy says it may be at least as important as keeping the club open."

"Last I spoke to Al, Tawny disappears in a couple of weeks. Does that still happen, even with all that's going on?"

"Yes"—Gooshie coaxed the answer from the small computer—"and Kyle never finds her again."

"This just gets better and better, doesn't it? No matter what I do to help Kat, we'll still fail, and the two people I probably *could* help want me to mind my own business. Real cinch, this Leap."

Gooshie continued punching the hand link. "Dr. Beckett, isn't Kyle Singer an investigative reporter?"

"So he says."

"Well, may I suggest...." Gooshie stopped, looking around. He took a breath that Sam could only describe as tentative. He raised an eyebrow and continued. "May I suggest that you enlist Kyle's help in exposing this councilman Barrenger?"

Sam stopped his pacing. "That's a thought. If nothing else, he may be able to tell us where to begin. It could also help me get them—Gooshie, what's wrong?" The programmer held his hands to his throat as if he were choking.

"Don't... know," he croaked. "No air...."

Sam's fears of a moment before came rushing back. "Another bug in the system?"

Gooshie was too busy punching in the door code to answer. He was turning as white as his clothes, and his eyes bulged. The hand link's blips and bleeps were distinctly dimmer. "Don't worry. Everything . . . okay," he managed before the glowing portal appeared next to him. He gave a quick wave and rushed into the light. The door closed behind him, severing the link.

Whether the wave meant "It'll be okay" or "Good luck, because you'll never see *me* again," Sam had no way of knowing. He prayed for the former, although the abruptness of the departure made him extremely uneasy.

Even if the system was out of air when he first entered the Imaging Chamber, Gooshie still should have had more than enough to last him for the short time they were in contact. It was as if the air were being deliberately pumped out.

Was system failure turning Ziggy psychotic? He supposed it was possible. If she had a personality, why couldn't she go off the deep end?

Even if that *was* the case, there was nothing he could do. That fact only added to the pall over his feelings. He tried to shake it off. It was past time to do what he could; maybe he could even kill two birds with one stone.

He joined Kat at the bar.

"Stomach okay, honey?" the woman asked, hurriedly putting out her cigarette. "You were gone an awful long time."

"Stomach's fine," Sam replied. "It's everything else that's going wrong." Kat looked at him questioningly. "What if I told you that Barrenger is going to win no matter how well we do on Tuesday?"

"I'd say you must have had one hell of a trip to the bathroom."

"All the same, he's rigged the election somehow. Just call it a hunch. He seems too calm, too cocky. Even for him."

"How can you be sure?"

"Indulge me. The first thing we have to do is get Tawny down here. We need to find her brother."

"That jerk?" Kat shook her head and walked to the other side of the bar. "Get my phone book out of the office."

By the time Sam returned, book in hand, a phone was on the bar. "Your instincts have been right so far," Kat said as she dialed, "so I'm not going to question them now. But I want . . . Tawny? Kat. Get your butt down to headquarters. And bring that brother of yours. . . . Well, I don't care. Find him. If you don't come now, don't bother coming back at all." She hung up and turned her mischievous look on Sam. "And people don't think I have what it takes to be mayor? Hah!"

"What's that you said about finding him?"

Kat furrowed her brow. "*Says* she hasn't seen him since last Saturday. And after the *stink* he made. Maybe he left town. Good riddance."

"No, I think we need him to pull this election off." Kyle *couldn't* be gone. Gooshie would have told him if that was the case. But, as he thought about it, Sam realized that Tawny had given him advance warning the night of the council meeting. He had been too emotionally wrapped up to hear her.

If Kyle hadn't contacted his sister since the night of their confrontation, and he hadn't left town, then what *had* he been doing for the last week?

"Honey, the day I need *him* for anything is the day I'll get into the ring and let you beat me in front of everyone," Kat said. "And, as you know, I don't take a beating easily."

Sam hit upon a possible explanation. It was a lot to hope for, maybe too much, but it also made the most sense. "Get a taste for humble pie," Sam replied. "I have a feeling you'll need it."

CHAPTER FOURTEEN

"You're going to need a hologram, Al."

Al nodded and shrugged more securely into the long lab coat, holding it closed in front. Not only did he feel naked, but the damned Fermi suit was bulging in all the wrong places. "I know. You're it."

Samantha looked up sharply from the console, her face awash in the chaotic color patterns it threw off. "Me? Why? What about Gooshie?"

"He'll be too busy here. We've never tried to pull off two separate Leaps at the same time. If there's a problem with the programming, and I can't afford to believe there *won't* be, then he's the best one to take care of it. The same goes for Tina and the hardware. Put together, they know more about the system than Sam does. *Or* you. They're too valuable on this end."

"Verbeena—"

"Will have her hands full with two Visitors."

Samantha returned to her study of the main console. Al didn't know if she bought his reasoning or not, nor did he care. The simple fact of the matter was that Sammy-Jo Fuller's genetic code made her the best candidate for a solid link. Where he was headed, a jumpy hologram could mean the difference between Sam's life and death. He wouldn't take the chance.

He couldn't tell her that, of course—not without opening up an entirely different can of worms. He had worked on the excuse since he made the decision to Leap. If her ego had been a little bruised because of it, he could live with that.

"There's something else," Samantha continued. "You've been avoiding it. How will we get you back? Dr. Beckett's retrieval formula is for the birds. You know that."

"I don't know." Al cast his gaze downward. He didn't have an answer during the impromptu committee meeting, and he hadn't come up with one since. "Maybe I'll just Leap back home."

"Yeah, and maybe they'll spring my grandmama from Peach Hill. But I doubt either is likely to happen. This is serious, Al. You'll be stranded in time just like Dr. Beckett."

"If you have another suggestion, I'm all ears." He was met with silence. "Look," he said more softly, "to be honest, the prospect terrifies me. But if I don't Leap and stop this thing, then what? Even though none of us would realize that we were living in an altered reality, I *do* know my options here and now. If I don't take a shot, Sam will die. And I can't let that happen. So I'll take the risk. If—*when*—when I stop Ann-Marie, we'll sort out the problems after. Sam comes first."

Samantha nodded. "Looking at it that way, I pity anyone who tries to keep me out of the Imaging Chamber."

"You're gonna love it." Al grinned. "You can walk through stuff and shoot your mouth off all you want. No one will see or hear what you do but me. And even if *I* get sore at you, I won't be able to do anything about it. You should see the games I play with Sam sometimes." He let out a low, desperate laugh. *Please, God, don't let it end like this*, he thought. He forced a brighter face. "Just one thing. When you first establish the link, you might get a little queasy."

"How so?"

"Ever have one of those Sit-and-Spins as a kid? It's a

little like that. You'll see what I mean soon enough. But otherwise, it's a real kick in the butt."

"Maybe that's what's taking Gooshie so long. Is he almost done in there?"

"He should be. Verbeena briefed him on his options before he left, and now that Ziggy has half a brain again, I don't think he should have many problems." Even so, Al wished things were moving a little faster. He had had enough of this hurry-up-and-wait routine. If they didn't start fishing soon, their bait would be cut for them. He raised his head and projected his voice. "Gee? Is it morning already? Where did I leave those wire cutters?"

"There's no reason to get cute, Admiral," Ziggy said serenely. "Using my 'half a brain' "—her voice dripped with animatronic contempt—"I believe I have pinpointed the divergence in the time stream. There is a ninety-seven percent probability that it occurs in the year 1988 at the Massachusetts Institute of Technology."

"MIT? But Sam left there in 1972."

"Dr. Beckett finished his studies in 1972, but he's not there as a student. He is attending a dinner being given by Professor Sebastian LoNigro in honor of his Nobel Prize. Professor LoNigro mentioned the upcoming event in an interview with *Time* magazine. The article appeared in the same issue that named Dr. Beckett the Man of the Year. It is my hypothesis that Ann-Marie Renerie obtained a copy of the magazine and is taking advantage of her foreknowledge of Dr. Beckett's location to commit the murder."

"So where do we go from here?"

"There are many people attending the function, Admiral. I suggest targeting your Leap to one of them."

"But who? I'll need to be able to move quickly. And Dr. Fuller will have to be able to lock onto me right away."

"How about campus security?" Samantha asked. "There's bound to be a few rent-a-cops there. That way, you'll be able to go anywhere on campus you may need to. You'll also be able to get backup."

"That's it," Al said, shrugging the lab coat from his shoulders. "It's showtime. Ziggy, get Gooshie out of the Imaging Chamber. I don't care if you have to pump the oxygen out of there to do it. I Leap now. Get Tina down here. Sammy-Jo, load the targeting program. I want to be head of security at that dinner. Got it?"

Samantha's fingers moved so furiously that she appeared to be massaging the multicolored unit. "Locked and loaded, Admiral." She punched the Enter code, moved to the front of the console, and stood erect. "Let's do it, Ziggy."

"Affirmative." A sheet of blue laser light shot from the opposite wall and ran down Samantha's form from head to toe. When it reached the floor, it arced up and repeated the process at a different angle. The light swooped up and down, up and down, circling the Control Room as if held by a gymnast in the midst of the trickiest of routines. "Three-dimensional imaging complete," Ziggy said as the laser completed its circuit. "Encoding the data into the Imaging Chamber's main control interface."

The Imaging Chamber's door opened and Gooshie stumbled down the ramp, rubbing his throat and gulping air.

Al rounded on him. "How's Sam?"

Gooshie's first attempt at an answer sounded like a cat choking on a hairball. He let a final hack fly in the admiral's direction, enveloping Al in a breath that could only have come from deep inside. *Well, I'm the one who ordered the air pumped out*, Al thought. *There's karma for you.*

"He appeared okay," Gooshie finally managed. "Of course, my perception could be called into question. The link wasn't very strong, and we couldn't talk much—"

Al cut him off with an exasperated sigh. "Is his life in danger?" He should know better than to ask the man an open-ended question by now. Gooshie was simply incapable of getting to the point quickly.

"No. No threats to his life."

"Does he have things under control?"

"We think so. He's hit a bit of a snag, but—"

"But you and Ziggy told him the best way to work it out," Al finished for him. "As long as he's still breathing and has all the options, that's the best we can do right now. Good work." He knew Gooshie had more to say, and it was not something Al could easily ignore, but he also knew he could trust Sam to fly solo for a little longer. Time for anything more was impossible. He turned, signaling the end of the discussion.

Tina approached from the opposite direction and went straight into Al's open arms. She kissed him long and hard. "Don't get hurt," she said when they were finished, "or I'll, like, kill you."

"Yeah." He hugged her tight. "I'll see you soon, okay? I love you." Although a bit husky, the words came out easier than they ever had. A real hell of a time to find the courage to say them. He let go reluctantly and turned back to Gooshie.

Al put a hand on the man's shoulder. Gooshie reciprocated the gesture; Al wondered if his face was really that easy to read. "Just take care of her, okay? Be good to her."

Gooshie nodded and drew Al into an embrace. "I won't have to. You'll be back in no time." Al turned toward the Accelerator Chamber.

"Calibrating Dr. Fuller's image to register with Admiral Calavicci's neurons and mesons," Ziggy said. "Hand link ready." The control panel on the console spit out a miniature twin that blinked in time. Samantha grabbed it, then stopped Al with a kiss on the cheek.

"I just wanted to do it while I was still able to," she said in answer to his surprised look. "It may take a little while, but I'll get you back. I swear."

"You'll do what you *can*," Al said emphatically. "That's all anyone can expect from themselves." He held up an admonishing finger. "And if I get wind of any more late-night pity parties in the cafeteria, I'll Leap back here for the express purpose of tanning your hide. I won't be held responsible for you losing your girlish figure. Got it?"

Samantha's expression shifted from consternation to hope in the course of Al's diatribe. She grasped his shoulders and shook him, causing her hand link to squeal in protest. "That's it! You're the one who's got it, Al. Don't go anywhere just yet." She laughed and rushed back to her position behind the main console. "Ziggy, access retrieval program AA2634."

Al looked on in confusion. "Just accept it, will you? Every retrieval program is a dud. We don't have time for more holdups. I Leap now." He resumed his trot to the Accelerator Chamber.

"Halt, Admiral!" Samantha yelled in a voice that stopped

Al instantly. His old drill sergeant wouldn't have been able to stand up to *that* tone. "Damn your macho heroism, Al. I think there's a way." She continued coaxing the control panel as she spoke. "That's the one, Ziggy. Load it up. What do you think?"

"Integrating program into my current parameters," Ziggy replied coolly.

"What the hell is going on here?" Al demanded, pacing restlessly.

"You're right in one respect," Samantha said. "Every retrieval program *is* a dud when we apply them to the problem of getting Dr. Beckett home. But you're *not* Dr. Beckett."

"Meaning?"

"The main reason for retrieval failure up until this point is that we can't get a good enough hold on Dr. Beckett's neuron and meson signature in the time stream. His constant blending with other people over the years has made it impossible.

"But if we apply my latest program to you—or any *fresh* Leaper, if you will—we may be able to get a fast enough fix to bring you back."

Al stopped pacing. He could feel hope attempt to creep in, but he stomped on it. "There's one thing you're forgetting; in Leaping, as in life, I'm no virgin. That lightning snafu a couple years back took care of that."

"It was an isolated incident," Samantha shot back. "And it happened so long ago that it may not matter."

"I wouldn't bet on it. I don't know if it counts as one Leap or two," Al went on, trying to recall the sequence of events. "It may even be *three*. It gets all magnafoozled in my mind." The primary memory was of the painful lump on his head. After that, it melted into fuzzy-edged frustration. He ticked off facts on his fingers. "First I had to switch places with Sam. Then I was Tom Jarret. Then Sam Leaped into me, trading places *again*. And I know for sure that we swapped some marbles during it all." He tapped his temple. "There's still a little bit of Sam with me. It's just sort of a feeling I have that was never there before." Not a very scientific explanation, Al knew, but it was easier than telling them he still heard faint echoes of choir song in the far reaches of his mind. "So, isolated incident or not, I wouldn't get my shorts bunched in excitement just yet."

"I would have to agree with the admiral on that point," Ziggy said, "though I have no shorts to bunch. The current update to the retrieval program gives us at maximum a sixty-two percent chance of locking onto his neuron and meson pattern. I cannot, however, accurately project how the time line divergence will affect those probability estimates."

"Damn!" Samantha pounded her fist on the console.

"Hell, I'll take those odds," Al said. "They're better than I would have allowed myself to believe. Now let's get moving. Get into the Imaging Chamber. Gooshie, Tina, take over here." He paused only long enough to fix the head programmer with a menacing stare. "And Gooshie, in light of the new odds, forget what I said before. If you go near her, I'll Leap again just to fix it so your mother never meets your father."

Gooshie gulped. "Affirmative."

Al ran to the Accelerator Chamber and punched the security code into the access panel. The rate of his heartbeat rose along with the door, and he entered the room, taking his place on the disk located in its center. "Ready, Gooshie," he shouted.

"Beginning countdown," came the faint reply. "Five, four. . . ." The voice was drowned out by the low hum that filled the octagonal chamber.

This is it, Calavicci.

The panels set at shoulder height in each wall began to glow with energy, casting Al in mellow blue relief. Atomic wind blew up from the perimeter of the disk, shrouding the chamber with a white fog that surrounded Al and crackled in his ears.

"Two. . . ."

The force of the blast doubled and the low hum rose to a shrieking whine. Al bid farewell to life as he knew it. His hair flew on end; his entire body tingled and his arms, suddenly very light, rose straight out to either side.

"Fire!"

The whine fell to a low rumble that was soon beyond his ability to hear. The atomic energy coursed through him, a roaring *whoosh* that carried him off in a blinding blue-white flash.

CHAPTER FIFTEEN

Sam's coffee steamed in the chilly morning air, but he didn't need it to wake up. Anticipation already had his nerves jangling like harp strings. He chewed impatiently at the rim of his Styrofoam cup.

"I still don't understand why I have to be here," Tawny said. Sam knew her cranky tone had little to do with being up early on a Saturday morning. "Whatever's going on is between you and him."

"Unless you come, the deal's off," Sam replied. "He made that very clear in our conversation last night."

"Please don't ruin this for me." Kat paced in front of the club. They had been waiting less than ten minutes, but she was already on her third cigarette. "Much as I hate to admit it, he might be onto something big. I need him." The words sounded smoother than Sam would have expected. Kat threw the cigarette down, and another immediately found its way between her lips. "I know the last thing you want is to see him, but consider it a favor to Richard and me."

"I didn't know I owed you any."

"Really? Who set you up with the security deposit on your apartment when you first got to town? And we didn't know if you'd stay more than a week." Kat's agitation was turning to ire. "If you want to be a little bitch, do it on your own time. For now, you owe me. So suck up and deal."

Tawny opened her mouth to reply, but Sam cut her off. "That's enough, you two. We have plenty to worry about without snapping at each other. So we'll take it for granted that you both apologize and chalk it up to a long night." Both women studied the ground, scowling. Sam resumed his gnawing and hoped Kyle arrived soon.

It *had* been a long night. Tawny hadn't gotten to the club until almost eleven and she was alone, claiming that she couldn't find Kyle. Sam didn't know whether to be happy about that or not.

When they finally reached him at his hotel at about one-thirty, Sam's hope became a bit more concrete. Not only was the reporter glad to hear from them, but he said he needed some help. Help with exactly *what*, he wouldn't say, but it had to do with Barrenger. He suggested they meet in the morning and discuss it.

So early Saturday found them waiting and wondering in the icy November air. Sam didn't know what Kyle had in mind, but for the first time in a week, he felt he was actively working toward a Leap out. He hoped the momentum wouldn't be cut short yet again.

All three heads turned at the sound of crunching gravel. A small orange car pulled into the lot. "That's him," Tawny said.

Kyle pulled alongside them and rolled down the window. "Good morning, ladies. Theresa."

"Still driving the ugly-mobile, I see."

"We haven't won the lottery since you took off," Kyle replied.

"Shall we?" Sam opened the back door and slid in. Kat pitched her cigarette and followed. Tawny stood in the lot until Kyle reached across the front seat and opened the passenger door from the inside.

"Still haven't gotten around to fixing that either, huh?"

She sat down and slammed the door. Kyle put the car in gear and pulled away.

"I've had my hands full." The reply implied volumes. Tawny looked at the scenery, pointedly ignoring her brother. A stony silence descended.

Sam purposely didn't break it. Despite his desire to know what Kyle was up to, part of his job was getting them to talk to each other. If that meant letting things fester until one of the siblings exploded, so be it. No one said the first step toward communication had to be pretty.

"Geez! I've been in livelier funeral processions," Kat said. "Now what's the story here? Where are we going?"

"Town Hall." Kyle shifted his focus to the rearview mirror as he spoke to them. "As for the story, I haven't exactly nailed it down. That's what I need you for."

"I knew it!" Sam exclaimed. "You're doing an investigative piece, aren't you? That's why you've been too busy to do what you came here to do."

"Smart *and* lethal. Not a bad combination . . . Sarah, was it?"

Tawny uttered a low laugh. "Just figures."

"What the hell is that supposed to mean?" Kyle snapped.

"Never mind."

"What's at Town Hall?" Kat insisted. Either she was oblivious to the unspoken argument or didn't care about it. Sam suspected the latter. "And why do you need us?"

"I have to get hold of some records and I've been having a problem. I think the woman in charge suspects that I suspect something. Every time I ask for the things I want, they seem to be in use. Or conveniently misplaced."

"What kind of records?" Sam's curiosity was piqued.

"They're the last pieces of a complicated puzzle." Kyle paused. If the way his eyes shifted from the mirror to the road and back was any indication, he was in the midst of some internal debate.

"You still haven't answered my question," Kat said irritably. "How do we fit into all this?"

"Look," Kyle finally managed, "I wish I could tell you more, but I promised a friend that I wouldn't reveal anything until I got my hands on some solid proof. It's not that I don't trust you; it's just a professional code I have to live by. My word means everything in my job, and I can't break it. Not for any reason. What I *can* tell you is that by helping me, you'll be helping yourself."

"I *told* you!" Sam snatched the unlit cigarette out of Kat's hand and threw it out the window. "Barrenger is fixing the election. How else could it involve us?"

Kat closed the pack and threw it into her purse. "Is that it?"

Kyle remained silent.

"What makes you think we can get this woman to be any more accommodating to us than she's been to you?" Sam asked. "She's bound to get suspicious if we're asking for the same information that you were, especially if she sees Kat."

"Kat won't be dealing with her," Kyle replied. "I will. And Theresa. Kat will provide a distraction. It's *your* job to get what I . . . *we* need."

"Me?"

Kyle nodded and turned the car onto Main Street. "You're the best choice. No one will give you a second look. I have a mole waiting on the inside."

"A what?"

"Someone who's on our team. They'll be waiting for you."

"How will I know who it is?"

"Don't worry, they'll find you." Kyle pulled into a parking spot in front of Town Hall. "They should be the only unoccupied person there."

"And what am *I* supposed to do?" Kat asked.

"Just be your usual flamboyant self," Kyle said with a smile. "Feel up to being the center of attention?"

"Always. Just point me in the right direction."

Kyle outlined his strategy. Sam was intrigued. He even got to use a secret password.

"It sounds okay," Sam said to the reporter. "But what if it doesn't work?"

Kyle fingered the faint bruise under his eye. "Then I'm counting on you to fight our way out."

• • •

Kyle escorted his sister up the granite steps and into the building. He turned to take a last look at Kat and Sarah waiting in the car, then nodded at Sarah's thumbs-up and let the door close behind him.

Once inside, Kyle detoured into the emergency stairwell and halted. He had told the others to give him twenty minutes before coming in. It would only take about five minutes to create the diversion. It was time to do what he had really come to this town to do. The fire door clicked shut with an echoing thud that filled the hollow space. Noticing that he wasn't following, Theresa stopped ascending the steps and turned. "Well?"

"Why, Theresa? Why did you leave?"

"If you don't know, then we have nothing to talk about."

"Yes, we do." He sat on the bottom step and patted one above him.

"I didn't come here to have a heart-to-heart, dear brother," she said sarcastically. "Now, are we going to do this thing?"

Kyle swallowed his response. "We have plenty of time," he said levelly. "Truth is, I didn't need your help with any of this. But insisting that you come along was the only way I could think of to get you to talk with me."

"One of your reporter tricks? An undercover sting operation?"

"Ambush interview, actually," he replied, indicating the step once again. "No demands this time. I just want to hear your side of it all. Now, please, sit. The clock is ticking."

"Since when do you care?"

"What kind of a stupid question is that? I *care*. I always have, even though you don't believe it for some reason. Why do you think I've been traipsing around looking for you all this time?"

"Because you're sick of doing your own laundry? You miss the cheap maid service? Wait, no, let me guess again. Dad has started shitting himself and you need someone to change his diapers?"

"Knock it off!" Controlling his temper was becoming difficult, but he would not let this degenerate into a shouting match. "Now either sit down and talk to me, or I get up and leave. If I do that, you'll never be bothered by me again. Of course, your friend will never have a shot at becoming mayor, either. So it's *your* call. I didn't want it to come down to this, but you're so damned stubborn...."

"*I'm* stubborn? You're the most obstinate person I've ever met!" She sat down to get a better vantage from which to yell at him. It was a small victory, but it was a start. "You've been that way since we were kids. You get an idea in your head of how things should be, then you expect reality to conform to it."

"That's ridiculous. Maybe I'm just more realistic than you are. And things seem to have worked out okay so far."

"Sure, for *you*. But what about me?"

"Last I saw, you were running away from your responsibilities and the two people in the world who really love you. But now that I see what you gave it up for...."

"Stuff it." She tried to stand, but he grabbed her wrist.

"Come *on*. Mud wrestling? Hell of a way to take charge of your own life. You couldn't find something respectable at least?"

"It's not like I'm a whore." Tawny twisted from his grip, but remained seated. "And it's good money. I bet I make more in a good night than you do in a week at that precious job of yours. And the hours are better. I figure if I stay at it another year, I'll have enough saved up to really do something. Take a couple of classes, maybe. That's an option I wouldn't have had at home, wherever that happens to be at the moment."

"*What*? What the hell do you think I'm knocking myself out for? Eventually I'll get a good enough job to put the family back on its feet again. Then you'll be able to have whatever you want."

"When you decide it's time? Don't do me any favors. Face it, you're not doing this for me, anyway. You're doing it for Mom. She's *dead*! But let's say you're really sincere. You're such a damn perfectionist that no job will ever be the right one. It'll always be wait, wait, wait. Before you know it, I'll be thirty-five and working in some supermarket in West Treetrunk, Iowa, with nothing to look forward to but the next step you'll make. Assuming you haven't burned out by then.''

"That's not how it'll be, and you know it.'' Kyle shook his head in frustration. "I swear you're being purposely difficult.''

"And you're being a selfish prick.'' When she stood this time, he didn't prevent it. How *dare* she? "Why don't you just tell the truth? I was right about Dad and you only want me to come back so you don't have to deal with the hassles anymore.''

"I admit you were right about him, okay? Happy? But tell me how I was supposed to know when he kept it from me.'' He glanced at his watch. They had to get moving soon. Five minutes, tops. This was not going at all the way he pictured it would. "The man just lost his wife, for God's sake. He was lost.''

"Don't give me that. He's a grown man. There's a time to let go of grief. He should have done it long ago.''

"He has. In a way, your leaving was the best thing that ever happened to him. He snapped out of it pretty quick when he ran out of clean underwear. But now it's time for us to be a family again. He misses you. And so do I.'' He stood to face her. "I know you're right about my perfectionism. Sometimes I wish I could be different, let go a little. But it's hard.''

Tawny sighed. "I'm afraid it'll be *too* hard.''

"All I can say is that I'm willing to try. Meet me halfway?'' He could tell by the way her eyes softened that he was beginning to reach her.

"I'll think about it.''

He nodded. "I suppose that's all I can ask for. Let's get moving.''

They went up the steps and found the hall of records. Kyle directed his sister to an adjoining room and sat her down at one of the microfilm stations. "Stay here. I'll be right back.''

He walked into the main area and located his target. He decided that the scowl on her face must be a trick of genetics; no one could be in such a bad mood all the time. "Lois, how are you today?'' He wore his most sincere grin.

The clerk stopped what she was doing at the sound of his

voice and looked up. "You again, Mr. Singer." She raised an eyebrow. "I expect you're here to make the usual requests?"

Kyle shook his head. "Maybe later. Right now I'm working on something different. Can you help me find some newspaper clippings? Since we've had some problems trying to get the documents for my original story idea, my editor wants a sidebar featuring Wilson's political history. I need to do some research."

The clerk jerked her head toward the room where he had left his sister. "Over there." She emerged from behind the counter and addressed the others who remained. "I have to assist this gentleman in the microfilm room, students. If you have any problems, I'll be inside."

"Did you call them students?" Kyle asked on the short walk back. Lois nodded.

"Part of the advanced classes in the high school. Civics course or community affairs or some such. Their best and brightest"—her voice dripped with contempt—"are assigned to volunteer at places run by the town. It's supposed to give them firsthand experience of how the municipal system really works." She snorted. "They just get underfoot, mostly. They're here only so they can pass their classes."

"Really?" Kyle asked in an innocent tone. "It's pretty interesting, but I know how you feel. My editor assigned a student intern to work with me today. She means well, but I don't think she has what it takes to make it as a reporter." He steered the woman toward where he had left Theresa. "I'm just glad that you're working today. Together, maybe we can prevent her from doing too much damage."

"Oh, you can be *sure* of that, Mr. Singer," Lois replied. "I'll watch her like a hawk."

That's exactly what I'm counting on, my dear.

Kyle made the introductions. Theresa played the part of a wide-eyed youth to the hilt. "We need a lot of stuff," Kyle said apologetically. "My editor wants a retrospective of all the mayoral elections for the last fifty years. We want to focus on the attitude of the electorate from the Cold War to the present." A bit too detailed for a runaround, maybe, but he was dealing with a meticulous person. "Do your records go back that far?"

Lois nodded. If he didn't know better he would have called the gesture enthusiastic. "They certainly do. And if you want some human input, I've been in this town for most of those elections."

Kyle laughed. The lure of the spotlight never failed. "Lois, that would be invaluable."

The clerk was off, somber skirt a-flurry. Kyle waited until she was out of sight before speaking. "You know what to do?"

Theresa gave a nod and smiled. "You really know your stuff, don't you?"

"Just part of the job," he replied. "If you can't get over on people, you'll get nowhere."

Lois was back in a blink, carrying a load of small boxes. "These are some from right after the war. World War Two, I mean."

"Terrific," Kyle said. "What do you say we get my assistant set up, then you and I can talk?"

"Sure." She produced a roll of microfilm from one of the boxes and moved to Theresa's side. "Have you ever used one of these machines, dear? It's very simple." She threaded the celluloid strip. "The image shows up on the screen. This button is to go forward, this one to go back. Got it?"

Theresa nodded. "I think so. Like this." She pressed the rewind button, backing the film out of the machine. "No, wait, I meant this one." She hit the forward button.

"No!" Lois yelled, but it was too late. The film shot forward, past the take-up reel and onto the floor in a pile. "Just look what you've done now!" She shot Kyle an exasperated look and was on her knees, gathering film.

Kyle looked back, rolling his eyes with an equally exasperated *see-what-we-have-to-put-up-with?* expression on his face. Theresa was beside the woman in an instant, all apologies, complicating things as much as she could.

Kyle glanced at his watch. Perfect timing. *Strike one*, he thought. *Now for two and three.*

Sam and Kat glanced into the records room. There was no sign of the older woman Kyle had described. When they heard a strangled cry from the adjacent room, Sam knew it was time. Kat nodded and walked into the main area.

"Listen up people," she said in a commanding tone. "I need some help here." She cornered three kids who suddenly didn't look so bored. "You, get me the town's business statutes. You, I want to see the petition guaranteeing my place on the election ballot. And *you* can get me the charter outlining the duties of town mayor." The kids looked on, shell-shocked. "Move it!" All three disappeared. Kat waved Sam in without looking.

Sam entered the room and looked for the one person who should be remaining. A girl stood behind a counter on the right side of the room. She was just hitting seventeen, and her dark brown hair was pulled back from her forehead in a red headband, just as Kyle described. It had to be her. He leaned on the counter. "Nevenka sent me," he said. "She said I should talk to Allison Wot ... Wata...."

"Wottawa," the girl finished for him. "Like the place in Canada, just add a 'w.' I'm Allison." She motioned for him to come behind the counter. "This way." She walked toward the back. Sam gave the room a final scan and followed.

Allison halted in a far corner and withdrew a manila folder from behind a filing cabinet. She held it out to Sam. "I think this is everything your friend asked for," she said. "It was a little tricky getting it all without Lois seeing." A mischievous glint lit her brown eyes. "But it was kind of fun, too. I hope it's all okay."

Sam leafed through the documents, his heart racing. He really had no idea what Kyle needed, but he was sure the reporter would have been specific. "They're just fine," he said, the words followed by a sudden laugh. The odds for a successful Leap might finally be tipping in his favor. "Thank you. You'll never know how much was riding on this. I probably owe you my life."

"Yeah, well, I owe you my A-plus. I have a feeling my journalism teacher is going to be impressed with my investigative work."

"You want to be a reporter?"

Allison shrugged with the blend of confidence and indifference common to teenagers. "Maybe. The course is pretty interesting, and if the job is like this all the time...."

"Allison, where are you?" The stern female voice was very near.

"That's Lois," Allison said. "I think it's time for you to get out of here." She led Sam to a back exit. "Just go through there. The stairway leads to the parking lot out back."

"You'll be okay?"

Allison chuckled. "If I can't handle Lois, then I deserve whatever's coming to me. Take off."

Sam gave her shoulder a squeeze and found the stairs. He was outside a moment later, walking through the alley to the front of the building, affecting the air of someone who belonged there. Kyle and the rest were waiting almost a block away, car running. Sam got in and they pulled away.

"Did you get it?" Kyle looked at him expectantly in the rearview mirror.

Sam held the folder up. "Got it."

Kyle let out a whoop and turned the car toward the club. "I got worried for a minute there. I thought that old bat was going to have a conniption with the way Theresa was screwing things up." He squeezed her knee. "Good job, Sis."

Instead of looking out the window, Sam noticed, Tawny smiled and squeezed her brother's hand back. He didn't know what it meant, but it was definitely a positive development. He prayed his turn of good luck wasn't just an illusion.

"I just wish you could have seen her when she spotted me," Kat added. "She got even more frantic. Oh, those poor kids! She ran through every one of them, checking out what they were getting for me. Seeing I had nothing you wanted threw her off the scent, I think." She made a grab for the folder. "Now let's see what the hell this is all about."

Kyle snatched it. "Sorry. I still can't fill you in yet."

"Just when I think you're not such an asshole after all, you prove me wrong," Kat barked. "I think we have a right to know, after what we did for you."

"Well, I'm a big proponent of everyone's right to know, but I still have to clear it with my source." He saw the annoyed looks all around him, even from Sam. "I'm sure they'll say it's fine to clue you in, but I have to make sure. Don't worry. Your answers are only a phone call away."

"Can you at least tell us if it will help us get around Barrenger's roadblock?" Sam asked.

"Darling, if I have everything here that I think I do, you'll be able to run right through it."

CHAPTER
SIXTEEN

Reality crashed back together one molecule at a time, and Al blinked away the afterimage of eternity. The first thing that struck him was that his shoes were too tight. He had traversed time and space, straight to an agony of snug leather and throbbing toes. Sam was right; God or Fate or Time or Whatever *did* have an odd sense of humor.

"Hello? Are you still there, honey?"

The voice came from a receiver pressed against his ear. He held it at arm's length and stared at it before fully realizing that he was in the middle of a conversation. He brought it back. "I'm here...."

"Which outfit was it you wanted me to wear tonight?" The voice on the other end wasn't entirely unpleasant and had an expectant quality. The tone made Al wish he had Leaped in about a minute ago. Or a few hours from now.

"Surprise me."

"Got it. I'll see you later."

"Yeah...." *I wish.* "See you later." The receiver clicked down on the other end and Al stood, staring down at dirty brown tiles in what looked like a vestibule of some kind, wondering what it was he would be missing. "Oh, *boy*."

A low rumble erupted from behind, and he turned around to face it. A sheet of brilliant white light rose to greet him and Sammy-Jo stepped out. "Dammit, Ziggy. Don't tell me we're having problems already!" Her image reminded Al of a time before satellites and cable television. "Admiral, am I getting through?"

"Barely. What's wrong with the transmission?"

Samantha looked up from the hand link. "It's the...." She looked questioningly at his shoulder. "What's going on?"

Al followed her gaze and realized that he was still holding the receiver at half-mast. Its cord dangled at his hip, ending in a tangle of twisted wires. "Geez! I Leaped into a conversation and was about to hang up when you got here. I guess I was more startled than I realized." He hung the useless unit back on its cradle. "What's wrong with you? Is the programming going ca-ca on us again?"

"It has nothing to do with the programming." Samantha poked at the hand link as if she was trying to jar it awake. "Time itself is playing tricks with us, near as I can tell. Ziggy says we're so close to the focal point of the altered time line that getting a firm fix will be next to impossible."

Al was glad he had chosen Samantha as his Observer. Had it been anyone else, he'd probably be on his own. "Well, at least we hit the target." He looked down at his clothes. Besides the tight shoes, he was wearing a nonregulation uniform. "Both targets," he amended, taking stock of his surroundings. On his right were doors leading into the drizzly night. He went left into a familiar-looking lobby, drawing the walkie-talkie from his belt. Samantha followed, flinching when the closing glass door passed through her.

Al felt a sense of déjà vu, but couldn't figure out why. He had stood in this exact spot before... kissing a girl? *Of course!* he thought, remembering dark eyes and even darker locks of hair entwined in his fingers. It was the place where he had first met that Lithuanian exchange student. Danesa! "This is the Stratton Center at MIT. It's the student union." He pressed down on the lever and held the walkie-talkie to his lips. "All stations, report." A hiss of static was his only reply. "All units, come in," he tried again. This time the static lasted only a moment.

"Dispatch. Identify."

"This is"—he looked down at his name tag—"this is Morgan over at Stratton."

"What's the problem, Morgan? Having trouble handling those astrophysicist thugs by yourself? Say the word and I'll call in the SWAT squad."

Damn! That could only mean he was working alone. "Uh, no. Call off the dogs. Things are just as exciting as you might expect. Just doing an equipment check. Morgan out." He didn't want to raise the alarm unless he had to. Lord knew what original histories he might disrupt by jumping the gun. Help was available if he needed it; that would do for the moment.

"Well, our dinner party is around here somewhere." Samantha shook the hand link. "Boy, is this thing a pain in the ass! Hasn't anyone ever thought to redesign it?"

"I think we did once already," Al said, trying to retrieve what hole-punched memories remained of his alma mater. "Didn't help one bit." He trotted the perimeter of the lobby. "I know there's a stairway around here that leads to a ballroom. . . ."

"Ziggy, reposition me." Samantha's holographic image vanished and reappeared across the vast area. "This way, Al. You take the long way. I'll see if Ziggy can center me on Dr. Beckett." She was gone in a blink.

Al took the steps down two at a time, each thud redefining his concept of pain. The sign greeting him at the bottom, Beckett Roast, was illustrated with a cartoon Sam spitted over a fire, apple in mouth. He chuckled at it despite his urgency; it had a familiar feel. He went in the direction the arrow pointed and soon found himself outside a large set of closed double doors.

A blowup of the *Time* cover was displayed on an easel in the lobby, vandalized almost beyond recognition. Sam's smile was missing a few teeth and he sported glasses, horns, and a goatee. It gave Al some apprehension until he saw the sign on top that read Booby Prize. Not Ann-Marie's work, then; just some good-natured ribbing to go along with the roast. Scrawled over it all was an inscription in handwriting that Al recognized: "All the best, Sam Beckett."

"Let's hope it stays that way, buddy," Al said to the photo. He checked his impulse to dash into the ballroom. Things were obviously still okay, considering how quiet it was. *Think like an assassin, Calavicci.* Another memory fell into place, and he looked for a side door.

The banquet room doubled as an amphitheater for some lesser student productions, in some of which he had had a part. There were catwalks and back corridors running the perimeter of the high ceiling that gave access from above. He had gotten to know the warren of nooks and crannies intimately during a production of *Cyrano* in his junior year. Danesa had given him Lithuanian lessons that had nothing to do with speech.

Anyone looking for a good perch could find it there.

"Oh, *shit*!" The door had been forced. Slivers of wood fell away from the latch as he pulled it open and rushed up the stairs. He made his way through the irregular darkness, halting in the swaths of light coming from the openings to the ballroom, purposely leaving his flashlight off. He would need surprise on his side.

A quick look down into the vast room told him that his original assumption was correct. The light conversation and laughter were interrupted only by the clinking of champagne glasses. His rush didn't give him time to spot Sam in the crowd, but he could see the head table clearly. If Ann-Marie was up here and had positioned herself directly opposite, a few turns in the corridor would bring him right to her. He moved quickly and quietly down the curving path.

The last turn loomed ahead and he stopped in a patch of darkness, slowing his breath. God willing, this might turn out to be easier than he thought. He reached for his gun . . . and realized campus security didn't carry. *Damn!*

He felt around frantically. Besides the walkie-talkie and flashlight, the only other thing hanging from his belt was a billy-club. He drew it up by the handle, twirling the tip to the crook of his elbow. A roundhouse swing would do nicely, but he would have to pounce if he was going to beat a bullet.

He steeled himself, body tense. Draw a breath. Hold. One . . . two . . . *three!*

He jumped around the corner, his momentum carrying him through the empty space to the floor. He landed hard on his knee with an audible crack, and his breath came out in a curse. She wasn't there!

He looked out into the ballroom to check his position. The head table lay below, directly across from him. If she was going to shoot Sam and get away with it, this was the only logical place. Then again, maybe rationality wasn't her strong suit. He moved to get up, but stopped in midpush. His hand pressed against something wet on the floor. He pulled it up and studied it in the diffuse light. A leaf. What better calling card for someone who'd been out on a rainy November night?

But where the hell was she?

Al was on his feet, running as fast as his damaged knee would allow, to the only other place she could have gone. Instead of doubling back, he continued in his original direction, completing the circuit of the shadowy corridor around the vast room.

The jog was incident-free, increasing his fears. What could she be planning? How close did she intend to get?

He came to another set of steps and barreled down, emerging on the opposite side of the ballroom from where he began. He remembered that this back corridor led to some storage and utility areas. Beyond was the emergency exit. Odds were she hadn't gone out that way; an alarm would have sounded. Al slipped into the ballroom.

He shadowed the perimeter as discreetly as he could, studying the clusters of people he passed. Sam wasn't among them. He didn't spot Ann-Marie, either. He stifled the panic that was rising. It was a big room. He would find one or the other.

Face after unfamiliar face greeted him all the way to the potted palm at one corner of the head table. Nothing. And there was nowhere else to go. He eyed the few people sitting on the long dais. A woman in a green dress gave him pause; no, not Ann-Marie, but he knew her. One of his old professors maybe? A bit too young, admittedly. . . .

A sudden squeal at his side startled him out of his study. "... miral. I've located th...." The fuzzy shape flashed out as quickly as it had appeared.

"Sammy-Jo? What's happening?"

Her image faded back, voice garbled. "... ro's office. She's on her way there n...." She was gone again. He waited a few beats, but she didn't return. Time was getting dangerously short.

What had she said about an office? Ro's office.... He went cold. LoNigro's office! Ann-Marie was going to LoNigro's office. That's where the murder was going to happen. He must have Leaped in just as Ann-Marie left the building. She had probably passed right behind Morgan on her way out, for Christ's sake!

Think think think! Where are the Physics Department offices? He recalled the Nuclear Accelerator Lab.... No, too far, all the way across the tracks. Nor was it a walk Sam and Sebastian would likely take on a blustery night. Where, then? It was a total blank. Damn his Swiss-cheesed brain!

He had to get moving. Maybe going outside would do him some good, jog his memory. He turned to leave, but a shrill beep drew his attention back to the main table. Sammy-Jo?

The beep resounded and Al looked around expectantly, but the Observer was nowhere to be seen. The woman in green provided the only motion, fishing her purse out from underneath her seat. With some rummaging she finally produced the culprit. The beeper in her palm continued its annoying song until she slapped it down on the table with a frown.

Al ran out the door with little idea of what to do besides keep moving. He took the blind turn toward the stairs at full speed and collided with the black and white mass that suddenly appeared in front of him, going down in a jarring tangle of arms and legs.

He rolled onto his back, wind gone, grasping his aching knee. He fought his way to a kneel and leaned over a man in a tuxedo who lay jumbled on the carpet next to him. "Can you possibly make this any harder?" he yelled in an upward direction. He turned the man over and fell back in shock. "You!"

He had collided with himself.

"What. . . . Where's my cigar?" His past incarnation fumbled one hand on the carpet and cradled his head in the other. "Why don't you watch where you're going, numbnutz?" He fixed a bleary stare on Al, his breath a sickening mixture of cigar stink and bourbon. His moving hand lighted on the smoldering stogie and put it back into his mouth.

Al felt as if the world had turned inside out. Years of forgotten shame, bitter hopelessness, came back to him, their presence a sudden, crushing weight.

It wasn't the first time he had been confronted with an earlier version of himself, but it was not something you easily got used to. And this wasn't the bright-eyed, physically fit Bingo of yore with his whole life ahead of him, either.

Al studied the rumpled tux, the three days' worth of whisker stubble, the alcoholic sway, finally stopping on the bloodshot eyes. The sight repulsed him.

"I'm talking to you, nozzle. What the hell is it with you people tonight? You could have killed me."

The drunk's nasty tone grated in his ears, turning Al's shock into anger. Rage crept in last, overcoming it all. "Killed you? I would have been doing you a favor. I doubt you would have felt it, anyway. How long you been on this bender? Three, four days?" He let out a disgusted chuckle. "It's one of the most important nights of your best friend's life, your only friend in the world at this point, and you don't respect him or yourself enough to show up sober."

"Who're you?" His double groused, suppressing a belch. He retrieved a flask from his jacket and unscrewed the top. "What the hell're you talking about?"

Al knocked the flask down the corridor, its amber liquid arching through the air, and grabbed himself by the shabby lapels. "You know exactly what I'm talking about, you pathetic little drunk!" Demons he thought he had conquered long ago cavorted in the double's bloodshot stare, and he fought the urge to throttle the man.

Memories of the evening were falling mercilessly into place. He couldn't remember exactly what he had said while stumbling around in front of the podium, but it was enough to cast a pall on the rest of the evening. "Sam is the only person left who hasn't turned his back on you, and you're going to let him down tonight because you're too sick and too selfish to think about anything but your next drink."

"I don't know who you think you are, but—"

"Shut up!" Al bellowed. "I've wasted enough of my life on you already! I don't have the time to start again. So here's your shot at redemption, your chance to give a little back. Where's LoNigro's office?"

"Wha . . . ?"

"Sebastian LoNigro's office! The Physics Department! I know you know him, so dig down into whatever brain cells might still be alive. You're not going to win, damn you! You hear me? Now which building?"

"Eastman, I think," the doppelgänger replied. "Eastman Labs in the Engineering Center. It's right across the street."

"Eastman! That's it." He let go and stood, looking down at himself.

"Who *are* you?" the drunk asked.

"Someone who knows exactly what you're going through," Al said more softly, "but refuses to pity you because of it." He turned to go, but stopped. "One last thing. It's something I've always wished I'd realized sooner instead of later, so listen. She's *gone*. She's gone for good. And you're not gonna find her in the bottom of any bottle. The sooner you get that through your sorry skull, the easier it will be. Trust me."

He started up the stairs, resisting the urge to look back. He couldn't decide if he wanted to cry or vomit. Maybe he had changed the original history of this horrible evening. That would be something, anyway. He would probably never know for sure, and it had nothing to do with the side effects of the Leap; his memory of the latter half of the 1980s was mostly a sickening blur of embarrassments between alcohol-induced blackouts.

Of course, none of it would matter unless he got to Ann-Marie before she got to Sam. He hobbled up the last few steps, putting almost all of his weight on the bannister. His knee was swelling, each bend meeting with a stronger protest.

He ignored the pain and focused on Sam. The thought carried him across the lobby and into the rainswept night.

CHAPTER SEVENTEEN

Ann-Marie peered into the night as she silently closed the door of the building. It wouldn't do for Beckett to hear her now, not when she was so close.

The only motion she saw through the moisture her breath left on the glass was leaves blowing across the walk. The guard wasn't following. Hell, he probably hadn't even noticed her, too busy slobbering into the phone. Ann-Marie shuddered at the image. He looked so much like Tibor that she was still debating whether to go back and kill him just for the hell of it.

No. She was here for a greater purpose, the greater prize.

The voices laughed in assent. They had been laughing for most of the last week, urging her on, confirming the justice of her actions.

Getting the gun had gone smoothly. Rodimer had come through with an impressive piece. She hefted its cold weight as she drew it from her waistband. Nice and heavy. She didn't know what kind it was, nor did she care. The only thing she needed to be sure of was that the bullets it fired would take most of Beckett's head with them.

Quit your dawdlin', girl! Justice is at hand. You won't be a victim no more, will you? The time for givin' in to fear is past.

Ann-Marie sprang into motion, stalking quickly and silently down the shadowy corridor, shrugging off her backpack as she went. She had taken her treasures with her, using them for inspiration. The magazine photo wasn't enough; her own drawings of Beckett seemed more real to her, more concrete. They also didn't impart the mysterious, dreadful fear she felt whenever she looked at the *Time* cover. But she would have to part with them for the time being; they would get in her way. A sloshing sound came from the bottom of the sack as it hit the floor. She unzipped the front compartment and took out the knife, leaving the rest. She would retrieve the sack of treasure later and add to it the last and best booty of all.

If she could find it. She had only tailed the two men closely enough to see them come into this building. She had no clue where they were now.

How hard can it be to find them, you weak-minded fool? They're the only ones here.

She nodded absently. Like shooting fish in a barrel. Search and destroy. Much easier than she would have thought possible. Dressed in jeans, flannel shirt, and preppie boots—looking like the typical New England college student searching for a quiet place to study—she hadn't had much trouble getting close to the arriving party guests or staying far enough away to prevent them from seeing the wrinkles around her eyes, the lines drawing her mouth down at the edges.

Her youth, her looks, her money—gone. Revenge was the only thing she had left, his death the only thing she could look forward to. And her luck in getting to him was holding.

When she had found that crawl space in the other building, she knew the gods were still on her side. Beckett had been in clear view below, circling the room from group to group, making the pathetic sheep bleat with laughter before moving on. Smug bastard. All she would have had to do was wait for him to stand at the podium to give his little speech, and *wham!* The part of him that everyone was making such a fuss over would be plastered on the wall, in clear sight for closer inspection. She couldn't have asked for anything more.

Beckett had talked for a while with an older guy who must be the professor named in the article. She had felt a moment of panic when the two had left the room, and had run back the way she had come. She stopped at the door, opened it a crack, and watched as the two men went upstairs to the lobby. They were leaving!

She had counted to twenty before concealing the gun and going after them. Staying discreet had become more important than ever. The gravel-voiced little troll had almost ruined everything, stumbling out of nowhere.

"Hey, doll face," she mimicked disgustedly into the empty corridor, "go back up the stairs and I'll give you a private lesson you won't soon forget." What the hell had he meant by *that*, the little weasel? She had repressed the urge to see how his cigar would look sticking out of his rump, and instead knocked him to the carpet before continuing her pursuit. He was so drunk he would eventually have gone down without her help. The weak-willed idiot probably couldn't recall his own name if asked; he would never remember her.

No, no immediate need to take care of him. Beckett first, then LoNigro, nice and neat. Maybe she would cap the professor first, if seeing him die would hurt Beckett. The options were thrilling! *Then* she would go back and kill the drunk and the guard. What the hell, she had bullets to spare.

Her circuit of the first floor was complete. Nothing. She found a flight of steps and climbed.

Come out, come out, wherever you are.

Thunder, Oshún's rage, boomed outside, adding to her internal rumblings. A lightning flash illuminated her first step into the corridor, spilling in from a window at the near end. Yemayá guiding her steps. Ann-Marie carefully surveyed the doors lining the hall. All were closed, except....

She walked deliberately toward the strip of light at the end of the corridor, Elegba's path to spiritual transfiguration.

The time is at hand, child.

The door stood slightly ajar, the brightness spilling from behind it a beacon, calling her, pulling her.
Strength, weakling.
Blood pounded in her ears. She could almost smell him.
"Your experiment is at an end."
She rapped the wood lightly with the gun, three times.

CHAPTER EIGHTEEN

"The party's back at Stratton, Sibby. What do you need to show me that can't wait until tomorrow?" Sam Beckett followed his mentor into the office, pulling the door behind him. The room smelled of polished leather and old parchment. Sam let it wash over him.

"That anxious to go to the slaughter, are you?" LoNigro asked, throwing his coat over a chair. He went behind the mahogany desk and pulled the bottom drawer open. "It won't be pretty, you know. They smile and congratulate you, but don't think there's a one of them who's not green inside." He straightened up, holding two champagne flutes by the stems. "I, for one, have labored all week on a limerick that will sear the skin from your bones. The string theory was tough, but it's nothing compared to finding a word that rhymes with Beckett."

Sam could think of three off the top of his head, but remained silent. He had the unfair advantage of having carried the name for thirty-four years. "What are those for?" he asked.

"These, my friend"—Sebastian set the glasses on the desk—"are for fifteen years' worth of sweat, sleepless nights, scrabbling for funding, and putting up with congressional focus groups." He gently pushed a wooden panel on the bottom of the bookcase; it opened on hidden hinges to reveal a mini-fridge. "Like it?" he asked, pulling the door open. He retrieved a bottle of champagne. "Just had them put it in last week. Think Hef would be proud?"

"All you need is a pipe and pajamas," Sam replied. "Very chic."

"And handy for keeping my Almond Joys cool. But, seriously, there's been a lot of limelight lately, and I haven't had the chance to congratulate you properly." He poured, and handed a foam-topped flute to Sam. "To you, Sam, and your Nobel. I'm proud of you."

Sam didn't raise his glass with Sebastian. He felt funny drinking a toast to himself. "The prize is as much yours as it is mine," he said. "It was in your cabin that *we* did the first calculations. And I know for a fact that without your continued input, I never would have made the numbers fit."

LoNigro shook his head. "You've come a long way from the Berkshires, my boy. Permit yourself an 'attaboy,' will you? You've done it."

"No, Sibby, listen. I've never really said this straight out, and it's past time. If you hadn't been there for me after my dad died, I don't . . . I don't know."

The memory was vivid, painful. His father leaving the world a broken man, the farm to which he had devoted his life stolen out from under him until the only thing left was the plot of land the house sat on. And worst of all, Sam being too self-absorbed to realize it until it was too late.

Sebastian had shown himself to be more than an interested physics professor during that terrible summer, taking Sam to his cabin to help him sort things out and presenting him with a theoretical exercise about the scientific nature of birth and death and the string that linked them. "You helped me put it into a context that I could understand," Sam continued. "I . . . I. . . ." The words didn't exist. "Thank you."

Sebastian shrugged. "Just simple mental gymnastics, that's all." They stared at one another for a few moments, sharing one of the expressive silences that constitute the meaningful memories in a man's life.

"On to more important things," LoNigro said, expression turning devilish. "What are you going to do with all the dough?"

Sam laughed and took a taste of the bubbly. "Get myself a lifetime pass on the gravy train and leave the pitiful rabble like yourself behind, of course. What do you *think* I'm going to do with it? Projects don't build themselves. I'll use some to set Mom up, but the rest is start-up capital for Quantum Leap."

Sebastian inhaled loudly through clenched teeth. "I think I just heard a few more of our esteemed elected officials run screaming into the night. They hate you more than our colleagues smiling away over in Stratton."

"They keep expressing interest in my theories," Sam said. "Weitzman, especially. Laugen, too. It's just tough to get them to make a commitment."

"Well, don't expect them to roll over for you now that you have a Nobel tucked under your belt. If anything, it will make them more obtuse. The only way they can still feel they have the upper hand is by proving to you that they're the ones with the power to make or break you."

"To hell with that. It's not my fault they don't understand the mathematical foundations. If I try to explain it intelligently, I get blank stares. If I dumb it down, they look at me like it's the most ridiculous thing they've ever heard. I'm getting tired of it. I'll go corporate if I have to. That's where the money is, anyway."

"To hell with *that*," Sebastian replied. "Do you really think they'll let you go your own way? None of them understand what in the world you're talking about, but it's for precisely that reason they'll keep stringing you along." He raised his hands. "No pun intended. Politics dissolved to the base, pure and simple. They'll doubt your sanity, they'll even doubt their *own* judgment for letting you present your theories time and again, but not *one* will wash their hands of you if there's even an off chance that you've stumbled upon the greatest discovery mankind has ever known."

"A bit melodramatic, don't you think? I mean, I wouldn't go *that* far."

"But anyone else would. You may have cracked time travel, for God's sake. The scientific community is spinning on its proverbial ear, my friend. You've got the unified field theory in a stranglehold, and that was only a by-product of reaching the greater goal. Never mind science, either. The government is salivating, thinking of the military applications. I told Tom, and I'll tell you again; you have the kind of brain that comes along only once in a generation. Don't try and shrug it off like it's no big deal. Uncle Sam won't."

Sam paced, draining his glass and loosening his bow tie. "Okay, point taken. The false starts are just getting frustrating. I didn't expect them after the work I did on Project Starbright. Still, I think the tide might turn soon. I've finally got someone to help me...."

"What? Who?"

Sam remained silent, mentally cursing himself. This was the one direction in which he didn't want the conversation to go, and he'd steered it there himself. Damn champagne!

"*Oh* no," LoNigro exclaimed. "Tell me it's not Calavicci."

"Sibby, he knows what he's doing...."

"The only thing he knows is how to get smashed," the professor said. "What the hell are you thinking? He'll bring you down in flames."

Sam turned, annoyed. "He knows how to avoid the pitfalls. He's spent his life dealing with these people, and he's not afraid to stand up to them when he has to. I need him."

"I know he's had a distinguished career, Sam, but it's over. They're only waiting for an excuse to bounce him. You're just taking on baggage you don't need because of friendship."

"That's not true. He managed Starbright just fine. He's hit a rough patch, I admit, but he just needs something to focus on to get back on track." Sam sighed. "I know it's a gamble, but it'll pay off. Just spend a few minutes alone with him and you'll see what a terrific guy he is. Not even the booze can hide that."

LoNigro shook his head. "You're making a big mistake,

Sam." There was a rap on the door, and the professor moved to answer. "Party calling...."

"Well, for once I'll prove you wrong. In fact, I'll bet on it." Sam put his glass on the desk, studying it absently. "What do you say, Sibby?"

Sam looked up and saw his friend backing into the office, hands raised.

Al rushed across Massachusetts Avenue, the damp air wrapping his injured knee with cold fingers. The layout of the campus was still fuzzy in his mind, but he didn't think about it. He had taken this route so many times in his youth that his subconscious would probably lead him there. Concentrating would only get in the way.

Which was just as well, because his mind reeled as it replayed the meeting he had just had. That it was unsettling was the mildest of observations; seeing himself at such a low point without the filters of time and impaired perception was a blow more crippling than his throbbing joint. He had been worse than he ever allowed himself to imagine, stumbling like a pantomime hobo, as irritating as poison ivy. The archetypal nasty boozer.

He had been sober going on twelve years now, except for one slipup he could remember, but a lifetime wouldn't provide enough distance.

Al wouldn't have thought it possible, but Sam had gained his respect from yet another angle. What had his friend seen in that wreck back there that had allowed him to risk his life's work, risk losing the fruition of his dreams?

He concentrated on what to do when he got to Sam. *If* he got to Sam at this hobbling pace. He didn't have a gun, so direct confrontation was out of the question. Sneaking up from behind would also be logistically impossible if they were already in LoNigro's office, which they most definitely would be by the time he arrived. He could lie in wait in the corridor until they came out....

And would probably wind up watching as Ann-Marie shot Sam and Sebastian. There was no way three live people were leaving that office unless he made sure Ann-Marie wasn't one of them. He was prepared to take that step, if it was necessary. He hoped it wouldn't be.

Surprise was really the only way, and if that meant rushing in and hoping for the best, then that's what he would do. Not that he particularly wanted to die, any more than he wanted to kill Ann-Marie, but he was also prepared to take *that* step if it meant saving Sam, and saving himself from a revived lifetime of the horror he had just seen in Stratton. No, the time line he had Leaped from was the better choice, even if preserving it meant he couldn't be a part of it any longer.

Al redoubled his pace, trying to outrun the leaves scurrying ahead of him in the gathering wind. The dark buildings surrounding him didn't look familiar, but they somehow felt right. If only he had a way to be sure. "I'll take any help you want to give me!" he yelled heavenward.

The clouds roared an ear-shattering response.

He caught a white flash ahead on his right, then lost it in the lightning bolt of the next instant. Had it been Sammy-Jo? He ran blindly toward the spot, but the only thing he saw was a path leading to the right.

Which way, dear God?

He decided on his original direction, but a familiar *ploink* sounded on his right after he had taken no more than one step. Unless the lightning strike had transported him into the Loony Toons universe, it could only mean one thing.

The new path brought him into a courtyard, and Al strained his ears for another signal from Ziggy. He wheeled around on his good leg, looking for something, anything.

Another bolt from the heavens revealed a sign. Eastman was directly in front of him. Whether it was blind luck or divine intervention, he didn't know, but he offered up thanks and sped into the building.

Once inside, caution slowed his pace. On the off chance Ann-Marie hadn't gotten upstairs yet, it wouldn't do for him to provide a target. He hugged the wall, just another shadow—providing, of course, that shadows tripped.

He went down, cursing himself, and came to a stop on his extended arms. He pushed himself to a crouch, his knee telling him in no uncertain terms exactly what it thought of *that*, and inspected the bundle at his feet.

A backpack. It could belong to a careless student, but he didn't think so. And if Ann-Marie had dropped it, it might contain something more useful than a nightstick. He unzipped it, but found only crumpled papers. He unfolded one to reveal an image of Sam as disturbing as some of his POW camp memories. He tossed it aside and dug deeper, toward the weight he felt in the bottom.

His hand cradled something smooth, and he drew it out. Not a gun, but an old mason jar filled with clear fluid. What use could she possibly have for that? He caught an acrid smell and raised the container to his nose. Formaldehyde?

He pushed off from his good leg, cradling the jar. It might come in handy, if only as a projectile. He didn't know Ann-Marie's intentions and tried not to figure them out. The sloshing sound coming from underneath his arm made it hard. The one thing he *did* know for sure from seeing those pictures was that Ann-Marie was beyond talking to. No speech he made, no matter how persuasive, would make her change her mind.

More thunder boomed from outside, and the first fat drops of rain splattered against the roof. He thought he saw Sammy-Jo come and go again, but he didn't need her to tell him what to do.

Al found the stairs and began climbing.

Ann-Marie jammed her gun into LoNigro's stomach, backing him into the room. It was funny the way he raised his arms, like a puppet on strings. *Her* strings.

"Who are you? What is the meaning of this?"

"Shut up." She rammed the barrel into his gut and he stumbled backward, falling over a chair.

"Sibby!"

It was Beckett! New hatred poured into the ball in her middle, and it puked forth waves of anger worthy of the years of frustration and wondering she had suffered, contracting her finger. She pointed low as she pulled the trigger, just missing Beckett's foot. He froze, looking at her.

"See what you made me do?" she seethed. "Someone

might have heard that." She leveled the gun at his head. "Just like you, screwing things up."

Fire, weakling, fire!

Her finger tightened on the trigger again, but she forced it to relax. "No!" She couldn't permit it to end so quickly. The gods had given her a chance to do more, to confront him face to face. To make him suffer.

Mete out his justice, child. The time is at hand.

"And if they were Thomas's murderers, would you give up the chance to make them understand? To know why they deserved to die?" There was no response.

"Sam. . . ." The professor was on his knees. Sibby, was it? What kind of sissy nickname was that? "Who is she talking to?"

"I told you to shut up, *Sibby*." Ann-Marie waved him to his feet, laughing at the ridiculous sounding word. "It's finally *my* turn to have some questions answered. Stand next to your friend, where I can see you." He moved at her urging, and both men backed against a large bookcase. What an obedient pair they made, linking their hands behind their heads without being told, elbows akimbo. A pair of frightened bookends.

She laughed again and relaxed the gun a bit, surveying the office. She was going to enjoy this. "Nice office, Sibby. I used to have one just as nice, maybe nicer. But your friend Sam knows all about that, don't you Sam?"

Beckett kept his eyes trained on her, opened his mouth.

"Keep quiet, dammit! You will *not* make me shoot you before I'm ready. My, but don't *you* look surprised." Ann-Marie chuckled. "Didn't think I'd find you, that it? Thought you'd get away with it and not have to own up to the harm you've done? Well, it doesn't work like that. I refuse to be your victim any longer. Do you hear me?"

"Uh, Miss, I—"

"I told you to keep quiet! It's your turn to know what it feels like to have your life ripped away when you're at your highest point. Mr. *Nobel*. Mr. Mirror." His blank stare enraged her.

"Knock off the bewildered act! Your little game is over. I know all about how you invaded my body like a parasite, destroying everything I worked so hard to achieve, all I sacrificed for." She felt tears welling in her eyes, but she refused to succumb to the weakness. He wouldn't get the satisfaction.

"What makes you think you have the right to screw with people's lives? With *my* life? And then toss me aside? Typical man. And you're honored for it! Well, I have your Nobel right here." Ann-Marie raised the gun again, moved in.

"It's time for payback, Mirror Man. I'm going to destroy *your* life, only I won't hide behind some fancy experiment and label it science. I'm not a coward like you. Pathetic weakling! Take a good look at me. I demand that you know exactly who it is that beat you. I'm the one you shouldn't have screwed with, and I'll be the last thing you see before you die." She was within arm's reach now, heart pumping, voice shrieking.

She knocked the other man on the side of the head with the barrel of the pistol. He reeled to the right, then collapsed beside the desk.

This moment was theirs. No one else would share it.

She rested the gun on Sam's temple, enjoying the way his breath quickened. "Scared?" She drew the knife out slowly, holding it in front of his eyes before trailing its point up and down his chest. "Good. Remember how it feels. Thelma will feel the same way soon. So will Katie." His breath stopped in a satisfying hitch.

"Did you think it ended with you? No, no, *no*, doctor. I know all about you. I'm going to erase every *trace* of your existence. Shouldn't be too hard with mother dear and sister living all by their lonesome out there in the sticks. It's too bad your father and brother are already dead. Only the pigs will hear their screams."

"How dare you—"

The pressure she applied to the blade shut him up nicely. Ann-Marie laughed. "My, you sound positively *enraged*. I'm glad you're taking it so personally. After all, what good is justice unless it stings a little?" She trailed the knife again, applying enough force to rend the fabric of his shirt, exposing the flesh underneath. "Don't worry. There is *one* part of you that I'll keep around. I feel it's my due, after all, since you took mine all those years ago. I only hope it's still beating while I cut it out."

You got him now, girl. If he ain't pantin' like a dog in heat....

"You always had a way with words, Evangelene," she chuckled. "But I've been remiss, haven't I? After all, this would be nothing more than a kangaroo court if I didn't give him a chance to come clean. Isn't that right, Sam? I can't go on and on about justice and not give you a chance to speak in your own defense. Speak up. I want to see you theorize your way out of this one."

"What are you talking about?" His facade didn't falter a bit. "How dare you threaten my family?"

"You got cotton in your ears, man? I just told you. Why do you insist on keeping up this charade? It's confession time. I'll give you one more chance."

"Chance at what? I've never seen you before in my life."

The ball in Ann-Marie's middle exploded like a supernova, the renewed rage causing her hand to tremble. How *dare* he?

The gods had made her choice plain. She cocked the pistol. "Wrong answer."

CHAPTER NINETEEN

The debate was on.

Sam paced behind the stage that had been set up, cursing his high heels and nervously rubbing his hands in the cool night breeze. Mayor McClough had decided to use the town square for the clash, partly to make room for the media circus, partly in hopes that the cold would deter people from coming. Row upon row of folding chairs stood under the large canopy, and most of them were already filled. The TV and press photographers' platform was jammed with people vying for space. Two reporters' areas—local and nonlocal—had been designated in front of the stage. Both were abuzz with the *whirrs* of rewinding tape recorders and the rustle of reporters' notebooks.

The townspeople weren't idle either. The sign-waving factions still marched the canopy's perimeter, a cacophony of conflicting slogans being chanted for any ear, human or electronic, that cared to listen.

Sam repeated a mantra of his own, growing more nervous with each step: "Where are you, Al?" He had no idea how things would end up, even with Kyle's information in hand. It was a foreign feeling, especially this late in the game. Usually he would know exactly what to expect at this point, but since there was no sensation of an impending Leap, there had to be a factor, maybe many factors, that he still needed to deal with. He racked his brain, looking at the situation from every angle.

Tawny and Kyle had spent the free moments of the last two days in quiet conversation. There had been no yelling, no abrupt partings that he had seen, but he wasn't sure if that was a good sign or not. For all he knew, they could be discussing Tawny's return home or just as easily be glossing over the issues that neither of them wanted to face, cementing the last bricks into the wall separating them. He wanted desperately to shoehorn his way into at least one of those bull sessions, to ensure the conversation was flowing in the right direction.

But he had his hands full helping Kat to organize a strategy, making sure she fully grasped the new information that had fallen into their laps. His apprehension sprang from the fact that *he* wouldn't be the one at the podium delivering the blows, knocking Barrenger off his horse. Ultimately, he had to trust his fate to her and hope she didn't screw up. Normally, he wouldn't have such a problem with a situation like this—Kat was obviously intelligent; would, in all likelihood, meet the challenge with ease—but without the Observer there to confirm it, he couldn't shake the antsy sensation. Being a man of science, of set questions and solutions, his faith wavered in the face of intangibles unless he was dealing with them personally.

Kyle's source was another problem. The person, whoever he or she was, refused to come forward. Kyle's permission to clue Sam and the rest of them in on the details was as far as it had gotten.

He couldn't blame the person for stepping lightly after he realized the magnitude of the situation. The possible repercussions were much bigger than Barrenger, would potentially rock the entire town. If Kyle's source was a town official, that person would become highly suspect for just having *known* about the situation. But understanding the source's hesitancy didn't make Sam feel any better. It was just another wild card to worry about.

What side of the story was he missing? What approach hadn't he thought of? He continued his pacing, praying he would soon hear the annoying melody of the hand link.

That was a side of the story he didn't *want* to think about. It had been three days since Gooshie's strange departure. The possibilities festered in the back of his mind, becoming more worrisome with each passing hour. First Al disappears, then Gooshie, as if the door to the Imaging Chamber opened onto bright white oblivion.

Even if he had something else to guide him, a crystal ball that showed him the outcome of the debate and exactly what he needed to do to ensure it, it would probably only make things worse. Then he would have nothing to distract his brain from churning out increasingly dire versions of the calamities befalling Project Quantum Leap. The only thing that could dispel them would be a haze of holographic cigar smoke and an inappropriate remark or two about the female anatomy.

"I'd swear you were the one about to step into the spotlight, the way you're hopping around."

Sam turned to face Mayor McClough, who had paused on the steps leading to the stage. "Just nervous, Ms. Mayor," he replied. "The club's employees have a lot riding on tonight. I hope everything works out."

"Well, that won't be apparent for a while, no matter who wins. You'll have your answer come this time tomorrow, when all the ballots are counted. The rest of us will just have to wait and see." She ascended the steps into a blaze of flashbulbs.

"I'm afraid we'll all be feeling change a lot sooner than you think," Sam said softly. He turned from the stage and saw Kat and Richard standing under a streetlight. Richard was speaking in soothing tones as his wife took drag after drag of her cigarette. Sam approached the pair. "Everything okay?"

"Just a case of the jitters," Richard replied with a smile. "You'll do fine, won't you honey?"

"Sure," Kat replied, exhaling a plume of smoke. "I've

been on stage before, in front of a lot *more* people and wearing a lot *less*." Sam wasn't sure who the comment was meant to reassure, especially the way Kat clutched the manila folder to her chest. She patted it. "It's in the bag. I just wish we knew exactly when to let the cat out."

"Kyle told us we would know. He seems to have everything under control, so try to relax. Just go up there and stick to our original strategy until it's time."

Kat nodded, taking another drag. "Got it."

Sam lingered a moment before moving off. He wished there was something more he could say, but he was afraid that if he kept talking, he would reveal his own doubts. Not a wise idea. He turned to the front of the stage instead and studied the frenzy of activity.

Mayor McClough had gotten to a live microphone and called for attention. The rallying crowds seemed torn between continuing their demonstrations and getting down to business. The groups grew smaller as people found their seats, most taking their signs with them.

The mayor stood in the center of the stage, flanked by a podium at each end, and waited for the crowd's rumblings to settle down before she started speaking. "First, I'd like to thank all of you attending tonight's debate. It gratifies me as mayor to see that the civic spirit is so alive and well in the community of Wilson. And since tonight's event has drawn so much attention outside our city, I'm going to establish certain ground rules that both candidates have agreed to.

"In an effort to keep the talk focused on the issues important to the members of this electorate, the local press will begin the questioning. They will have up to thirty minutes to ask questions before the floor is opened for townspeople. When *their* half-hour is up, the out-of-town and national news organizations can begin." Discontented rumblings arose from the national press corps. "Sorry, folks, but I will *not* permit sensationalism to detract from the real reason we're here." The comment was met with silence.

McClough continued, "Questions should be addressed to a specific candidate. His or her opponent will have a chance to respond. I will call on reporters one at a time.

"Due to the highly publicized nature of this election, the

candidates have agreed to limit their opening statements to one minute. And finally, I ask those people here tonight in a show of support to hold their applause until after their candidate speaks. Save it for the cameras, folks. Without further delay, I'd like to welcome Mr. Rex Barrenger and Ms. Kathy Scherber-Danson to the stage.''

The crowd erupted in applause as Sam walked along the edge of the canopy, looking for a seat. His heart leaped into triple-time when he saw Kyle sitting with Tawny. The journalist waved him over, indicating an empty chair next to him.

"Isn't this terrific?" Kyle asked when Sam was within earshot. He had to yell over the crowd's roar. "And we haven't even gotten to the good stuff yet!"

"What the hell are you doing back here?" Sam shouted. "Why aren't you up there getting ready?"

"I'm out of town, remember?" Kyle drew Sam down into the empty seat. "I'd have to wait at least an hour before I could even *think* about asking a question. But look at this!" He swept his arm in the direction of the crowd. "With all these people here, it's gonna be one hell of a show."

Sam couldn't share the man's exuberance. "I hope you're right." He watched Kat take her place on the stage. The woman looked like she was trying to hide behind the podium. "Can you at least give me a hint as to what's going to happen?"

"You'll know soon enough," Kyle replied.

Sam cast a pained gaze at Tawny.

"You don't have to tell *me* what a pain he is," she said. "Just be thankful you haven't had to put up with it your whole life."

Tawny's smile told Sam there was no real sting behind her words. Maybe those chat sessions had been positive after all. She also didn't seem worried. Whether she was resigned to fate or she trusted her brother a great deal, Sam didn't know. He just hoped her serenity didn't stem from the fact that she would soon be gone.

The tumult started to die down as members of the crowd took their seats. Barrenger basked in the applause, hanging on to every last shout and whistle. Kat looked positively ill. The way her hands fidgeted, Sam could tell she needed a cigarette—so much that his lungs ached in sympathy.

McClough turned to the councilman. "Mr. Barrenger, you won the toss, so please begin."

Barrenger took a few calm breaths, surveying the crowd, seemingly making eye contact with each person in the multitude. "Thank you, Madam Mayor. As you have pointed out, this election has become highly publicized in past days. I don't know, I guess it makes for good television. The only thing *I* have seen in these so-called news reports is myself cast as the heavy, the evil, narrow-minded fiend who wants to stop constitutional freedoms and do away with civil liberties. Let me set the record straight on a few points.

"Those of you living in this town know my record of service. I have always had the best interests of this community at heart, and I always will. I am confident that you good people will take *that* into account when you're at the polls tomorrow. I also feel I've made it abundantly apparent that I believe in First Amendment freedoms. In the face of this media barrage, I have answered all of the questions put to me by these news shows in a calm, succinct, and professional manner. And I've responded *only* to those questions that focused on the issues pertinent to the voters. Never *once* have I shown up for an interview in a bikini."

Barrenger stood back, modestly oblivious to the applause that welled up from his supporters. Rural Purity League signs waved about the crowd like albino butterflies. The mayor had to clear her throat several times to regain silence. "Ms. Danson?"

Kat stood straighter and stared at the microphone like it was a rattlesnake ready to strike.

Come on, Sam thought. *Just like we rehearsed it.*

"I guess we can all be thankful for *that*, Mr. Barrenger." Kat's voice shook, but the joke got laughs. "But I have not come here tonight to trade barbs. I've come to trade ideas, to discuss my vision for the future of Wilson. As I've said in any number of interviews, I didn't get into this election for the sole purpose of keeping my club open. If *that* was my goal, I could have packed up and gone elsewhere. I *certainly* wouldn't have put myself in the running for a new job.

"I'm running because I have come to realize how much

Wilson means to me. It has a progressive and forward-looking environment that should be a model for other towns to follow. And I don't deny that you had a lot to do with making it that way. You should be commended for the work you've done. But you've crossed a line. By deciding what is right and wrong for people on the basis of your own prejudices, you are undoing all you have achieved.

"That is why I choose to run. That's why I'm standing here now. And as for my interview technique"—her gaze swept the crowd—"it got your attention, didn't it? If I don't win, and Mr. Barrenger restricts your freedoms further, at least none of you can say you didn't see it coming."

Applause rose under the canopy once again, and this time Sam joined in. For a nervous moment he thought it would all be over before it even began. But Kat had segued into her rehearsed speech just fine. If only she hadn't delivered that last, impromptu, comment. It had gotten as good a reaction as the first, but its chiding tone might have been a bit much. He hoped she would stick to the prepared answers until her opening came.

"Let's move on," Mayor McClough said, positioning herself in front of the local press. She picked randomly from the waving hands in the front row. A runner took a microphone to the reporter who stood.

"Jerry Browne, NewsTeam Ten."

"Oh, *this* should be good," Kyle said under his breath.

"Is this it?" Sam asked expectantly.

"Only when I join the Legion of Morons. That little...." Kyle's voice trailed off into a string of mumbled invectives that Sam tried to ignore.

"Mr. Barrenger," Browne said, "what do you see as the main concerns facing the town of Wilson as we head into the next decade?"

The councilman barely considered the question before replying. "Chiefly, it will be maintaining our high standard of living. The city council has worked hard with the current administration to get the ball rolling, but it won't do us much good in the long run if it's not rolling on stable ground. Our revenue base must remain firm, and the way to do that is to implement long-range programs that will ensure employment in the years ahead. We can do that by extending a friendly hand to both industrial and professional interests. Tax incentives are one way. So are enterprise zones." He looked at Kat. "Ms. Danson?"

Sam's heart sank, the sensation a reflection of Kat's expression. The answer they had prepared was almost the same—if not in words, then in tone. He hoped Kat would be able to come up with a different angle that was positive and original enough to stand out.

Kat stood dumb for a moment before responding. It made her appear not to have been listening. "Well, Mr. Barrenger makes some good points. I . . ." she paused, her voice wavering more than it had before. "I think he's summed it up pretty well."

It took a few moments of silence for Sam to realize that she was finished. "She's choking," he said. "We're finished." He ran a nervous hand through his hair. He suddenly felt very warm.

Kyle put a hand on his knee. "It's not over yet. Just be patient."

Sam nodded absently, only half registering the fact that Kyle didn't remove his hand after the friendly pat.

McClough was just as surprised. "Next question," she said hurriedly, picking another reporter out of the crowd.

"Thank you, Mayor. Beau Stevens, Channel Eight. Ms. Danson, so far a lot of attention has been focused on the concerns of the city proper, but there are also many outlying areas that will be affected by your leadership. What do you have to offer to the farmers living on the outskirts of town?"

Sam kept his fingers crossed. He *knew* this would eventually come up; such matters had once touched him personally. She had this one down cold.

"It's plain that the farming crisis shows no sign of letting up," Kat said with an enthusiasm bordering on overcompensation, "and subsidy checks can only go so far. I'm not saying that we can solve the problem overnight, but we *can* offer alternatives to those struggling to remain independent. Offering continuing education courses in business management and technological advances at the local level is a good way to start.

"If elected, I will also establish a committee composed of people from area farms to work in conjunction with a representative of the Department of Agriculture to help their neighbors make the right choices in everything from crop planting to livestock management from one season to the next. This strategy of collective support from friends and neighbors is the strongest weapon we have. I say let's use it."

"Ms. Danson's sentiments are noble, to be sure," Barrenger said, "but I don't know how well they'll translate into reality. Committees are political animals, and if there's one thing I'm sure the town's farmers have had enough of, it's political solutions. The ones offered by the federal government are testament to that. I feel *I* can approach this problem on a more personal level, since my formative years were spent on my family's farm. I know what it's like to see your crops wither, to see an entire year's work going to seed. And I think that enables me to offer more than paper solutions. In my administration, I would do more than implement arbitrary programs and hope for the best. We've had enough of working *with* the problem. I will work *against* it by attacking it at its heart. The only committee I would form would be one to crusade for change in Washington. That is the type of long-term thinking to which I referred before. It's the only way we'll ever see real change."

Sam buried his head in his hands. *Strike two. Better get used to this body, buddy boy. . . .*

The next reporter was chosen. "Nancy Dotorovic, NewsTeam Ten."

Kyle's hand tightened on Sam's knee. Sam looked at it like it was some strange animal and shot the man a questioning glance before removing it.

Thank you! He directed his thought to the forces above with a roll of his eyes. *This is* just *what I need.*

"Mr. Barrenger," Dotorovic said, "can you tell me what knowledge you have of Southside Carting Company?"

Sam drew his attention back to the action just in time to see a flash of bewilderment cross Barrenger's face. The councilman recovered so quickly that most people probably missed it, but to Sam it was like a flicker of daylight in a dark tunnel.

"I seem to recall that name having come up in council business recently, but I fail to see where it's relevant to tonight's debate...."

Kat took advantage of the lull. "*Really*, Mr. Barrenger? Considering what I've found out about that company, I'd call the question *very* relevant." Kat turned to the crowd, her natural flamboyance finally emerging as she studied the upturned faces. "It's a bit complicated, so I hope you'll all bear with me as I start from the beginning."

"Now, Ms. Danson," Barrenger interrupted, "I thought we had agreed to stick to topics important to the voters."

"Mayor McClough," Kat said, "I believe it is my right to respond to this question without interruptions?"

"Those are the rules we agreed upon. Please hold further comments, Mr. Barrenger."

"Yeah, Rex, can it." Kat turned back to the crowd. "Now where was I? Oh yes, at the beginning. Let's go back to the *very* beginning. Mr. Barrenger launched his mayoral campaign with funds provided by an organization called the Rural Purity League, an organization that has gained a lot of support in recent weeks.

"It might interest you to know that at its base, the RPL is officially composed of only two men. One is a land developer and the other manages the Southside Carting Company. Neither is a man you'd think would have any particular stake in maintaining rural purity—certainly not men you'd think would be concerned over whether or not a bar on the outskirts of town stays open. Unless, of course, you take into account the facts that we have uncovered."

Kat held up the manila folder. "I have here documents that were retrieved from Town Hall recently. They are very interesting, and I'd like to share them with you now." Kat made a show of folding back the cover and held up the first page on the pile.

"On file was this contract. The deal it concerns is a simple one. Mr. Barrenger has hired the developer I mentioned— one of the heads of the RPL—to level a tract of land on the outskirts of town. It just so happens that my club sits on said land." Kat retrieved another document.

"*This* is the business charter for the Southside Carting Company. While the other man I mentioned may *manage* Southside, this paper says that it is, in fact, Mr. Barrenger who holds controlling interest in the company. His affiliation is of a silent nature, however." Kat picked up a new sheet and held it beside the charter. "Now *this* . . . I have to assume Mr. Barrenger has a very good friend working at the hall of records, because *this* is a bid that was on file, filled out by the manager of Southside and signed by Mr. Barrenger, to haul the town's trash. Interesting, when you consider that the town hasn't even called for such bids. At least not *yet*. I can only guess that Southside had this lying in wait, just in case. I'd say they have a pretty good chance of getting the job, too, since the form has been filled out in pencil." Kat lowered the sheets, but stopped and waved the business charter again. "By the by, this paper *also* lists the accounting firm that handles the company's finances. Just tuck that nugget away for now. I promise I'll get back to it."

Sam laughed to himself, clapping Kyle on the shoulder. Kat was back in form, a commanding figure whose words everyone hung on.

"Now for the capper." Kat held up a stack of paper. "Here I have the minutes from every town council meeting in the last year. Peppered throughout these documents, you will find evidence that Mr. Barrenger has been pushing through rezoning proposals, systematically designating the land around my club for heavy commercial use."

McClough crossed the stage to Kat, putting on a pair of glasses as she did so. Kat stepped aside as the mayor inspected the documents that had been held up. "Ms. Danson, exactly what allegations are you making?"

"None, your honor. I'm simply bringing to light facts that I find odd." Kat studied the crowd. "But I *guess* that if you wanted to look at it in those terms, you might say that Mr. Barrenger's crusade to shut down my bar was self-serving, in that he would be able to clear the land if he drove me out. On another level, it was a good distraction for both the town and the council, making mundane business like rezoning and development even more mundane...."

"She's playing it perfectly," Sam said in a low voice to Kyle. "We've won."

"Not yet...."

Kat stood by as the mayor sorted through the pages. "Oh, yes," she said. "The accounting firm I mentioned before? You will see that it's the same accounting firm Mr. Barrenger pushed the town council to hire to tally the election ballots."

"Is that *so*?" McClough held the papers in front of her and turned to Barrenger, shooting him a questioning look over the top of her glasses. "What do you have to say about all this, Mr. Barrenger?"

Barrenger stared for a moment, mouth working. "Madam, I must protest," he managed. "We have come here to discuss our respective visions for the future of this town. Instead, Ms. Danson bogs the proceedings down with a stream of nonsense."

"If that's what you *want*," Kat said, "I think I can detail the future as *you* would like it to be, Mr. Barrenger. You said earlier that the town must extend a friendly hand to industrial interests in order to survive. How long were you going to wait after the election to unveil your plans for the town's new landfill?"

"Ms. Danson, your allegations are unfounded...."

"How long before you started raking in the profits from your carting firm?"

"... and preposterous. The audacity you have displayed simply by participating in this race has bordered on mockery from the start. Now you dare sully it further with this dog and pony show—"

"*You* speak of audacity?" Kat turned to the crowd. "I ask you people, who has more audacity? Mr. Barrenger has abused the power of his council seat to railroad his personal agenda into legislation. It was your tax dollars he would have stolen to develop the land, your tax dollars that would have paid Southside Carting. *Your* money in *his* pocket, that's his vision for the future of this town."

"Once again we see the depths to which you will sink to make sure your snake pit remains open—"

"Enough! Both of you." McClough silenced the pair. "Ms. Danson, the implications of what you claim to have found are serious. Can you substantiate the authenticity of these papers?"

"She can't, but I can." Allison Wottawa appeared stage left, carrying the runner's microphone.

"Who are you?" the mayor asked.

"Allison Wottawa. I'm in the advanced program at the high school, and I intern at the hall of records. *I* found those documents and gathered them for Ms. Danson. She didn't make them up."

"This is ludicrous!" Barrenger gripped the podium like it was the only thing keeping him off the floor. "I will not stand here and have my credibility questioned by a child and a whore." The councilman wiped a sleeve across his forehead, hand shaking. "You . . . you have sunk to a new low, Danson, calling a child into service to perpetuate the wickedness. . . ." Barrenger trailed off, breathing deeply.

"It's over, Barrenger," Kat said calmly. "You've used me as a red herring long enough. Instead of calling on the people of Wilson to further support your corruption, I suggest you call your lawyer."

"I suggest you do the same, Ms. Danson," McClough said. "I think questions are in order, and I'll need answers from both of you." She addressed the crowd. "This debate is at an end, folks."

The mayor's words acted like a pin on the silent bubble of tension that had formed under the canopy; people erupted from their seats in a frenzy of shouts, and the reporters shouted questions at those leaving the stage.

Kyle laughed in triumph, jumping to his feet. "Out of the mouths of babes," he exulted, fixing Sam with a cheery smile. "*Now* it's over!"

CHAPTER TWENTY

Al focused on the pool of light at the end of the hall. A crazed female laugh punctuated his slow approach, the sound chilling his marrow. But it offered hope at the same time. If Ann-Marie was distracted, he just might get his chance. He kept his back to the wall and shuffled along.

He slid the mason jar from his armpit to his left hand and gripped the billy club harder in his right, lamenting his ridiculous arsenal. It was worse than having nothing; at least nothing wouldn't offer the illusion of hope. He would make it work, some way. He had to. He reached the door frame and held his breath.

A metallic click broke the silence, the unmistakable sound of a pistol being cocked. He forgot about strategy and swung into the doorway. Ann-Marie held a gun pressed against Sam's head. She shook all over.

"Sam!" Al cursed himself even as the word left his mouth. So much for surprise. Ann-Marie wheeled around, gun raised, but Al was already in motion. The mason jar flew awkwardly from his fingers in an overhand lob and he dived for the cover of the desk. A shot exploded in the small space and Al heard the jar shatter, smelled chemicals, and felt a white-hot sledgehammer crush his injured leg. The force of the bullet sent his airborne body into a half turn before he crashed to the floor beside the desk.

His knee thumped leadenly on the thin carpeting, and he cried out in pain. Silver flecks danced at the edges of his field of vision. But he dragged himself across the floor with both arms, retreating further behind the mahogany barrier. He saw Sebastian LoNigro on the other side, struggling to sit up. Blood ran down the right side of the professor's face from a nasty gash in his temple.

Al squirmed toward the man, gripping the lapels of LoNigro's tux to pull himself to a sitting position. The motion helped shake LoNigro out of his daze and he tried to stand. Al held on. "Stay down!"

"No. Up, both of you."

Ann-Marie loomed over them, staring down the barrel of the gun. She jerked it in an up and down motion. The men struggled to their feet.

"Stand next to him," she said, backing toward the door.

Al got an arm around LoNigro and they stumbled toward Sam. *Really* Sam! Not just his physical aura surrounding someone else's body; not just a hologram. His friend, living and breathing. For a moment, Al forgot about Ann-Marie, forgot about broken time lines, and reached for the man. For the first time in five years hand met hand, and Al gripped tight.

"You okay?" Sam asked, putting a supporting arm around Al's back.

Al steadied himself. "Been better." He tested his knee, but the pain made him dizzy and he quickly took the pressure off. As near as he could tell, the bullet had hit him in the thigh. But the entire leg was aflame, knee stiff beyond bending.

Ann-Marie reached the door and slammed it shut, kicking it once for good measure. "No more surprises." Her fevered gaze turned to Al. "I didn't think you had it in you, fat boy." She looked at the floor, eyes suddenly distant. "He *does* look like Tibor. I told you so!" Her eyes snapped back to Al.

"Kind of glad to see you, actually. It saves me a trip."

Al hopped on his good leg, fighting the lightness in his head. What did she mean by that? "You won't get away with it."

"And I guess you're here to see to that?" She bent and retrieved his nightstick. "You were going to save the day with *this*?" She laughed as she pulled the trigger. Slivers of wood flew from the corner of the desk.

Al tossed around possibilities in his mind. How could he gain the upper hand? He kept his face calm as he nodded toward the shattered jar. "And that."

Ann-Marie's eyes fell on the broken glass and she yelled, throwing the club at the pile of shards. "Look what you did! *Now* what am I supposed to carry it in? Just for that, I'm going to kill you first." She raised the gun.

Al reached through his terror and somehow managed to chuckle. "Why not? First a smuggler, now a murderer," he said softly. "What's next, Ann-Marie?"

"What? How do you know...."

"Know what, your name? You're not as mysterious as you'd like to believe. I know all about you." Al struggled to keep his voice even, to keep his face stoic, as the barrel of the gun jumped from him to Sam to Sebastian and back. He fought to remember Sam's Leap into Ann-Marie, what Verbeena had told him about her—anything that might knock the woman off balance. "Your name is Ann-Marie Renerie. You're from upstate New York and your father was a doctor. You got your degree in art history."

"Who are you?" Ann-Marie's voice wavered, as did her aim.

Al hopped on his good leg, pretending to shift his balance as he pulled forward. The other men realized what he was doing and moved with him. They were a step closer. "You went to New York City in 1972 and got a job at an import-export company run by Jon Colton." He jumped again. Ann-Marie stared, mouth open. If he could keep her listening long enough for all three of them to get close enough.... "Only it wasn't really an import-export firm, was it?" Ann-Marie stood dumb. "*Was* it?" he roared.

"No," Ann-Marie mumbled, eyes distant.

"No. And within four years you were practically running the operation. Smuggling. Robbery. Extortion." He hurled the words like shots of contempt. Each one drove the barrel of the gun a bit lower, made her mind retreat further, until the weapon dangled at her side. He had obviously found the right buttons to push. If he could find a few more of them, he would gain total control. "But it wasn't enough was it, the fancy office"—a hop closer—"the power? None of it was big enough to hide behind. All it did was give the fear sharper edges." Al moved as quickly as he dared, his words designed to go for her throat. "All it did was land you in prison."

Ann-Marie snapped her head up, eyes wide. "I'm not going back there. I wasn't supposed to go there in the *first* place!" Realization flooded her face, and she stared at Al as if seeing him for the first time. "You're with *him*!" she spat. "That's the only way you can know all these things. You're mixed up in his experiment!" The gun was up again, and all three men froze.

"Experiment?" Al asked impatiently. "Now that the power is gone, you have to hide behind conspiracies? Why are you so unwilling to own up?"

"No! You're not going to trick me! I'm not the one who has to own up. *He* is!" Her body quivered as her eyes darted from one face to the other. "It's him that needs to take responsibility for what he's done to me."

"I mean owning up to *who you are*!" Al yelled. "How you handle what life gives you. You can't do the wrong things and pin it on someone else." Al could tell he'd found another tender spot. Ann-Marie kept the gun raised, but her mouth worked in a stream of mumbles. They weren't directed at him, apparently.

"But you said justice was at hand," Ann-Marie stammered. "I *can't* be remembering wrong, he's standing right here. It's the one in the dreams. Him . . . the mirror." Her trembling thumb pulled back the firing pin. "Elegba showed me."

Al stared at her finger and wondered if he had gone too far. She repeatedly rested the digit on the trigger, then lifted it off again, clearly in the midst of some internal struggle.

The constant chatter accompanied her movements, and he heard the words "mirror" and "blue room" a few more times. She clearly *did* remember the Project, but couldn't figure out how to make it fit. And who was Evangelene? No matter. Get while the gettin's good, he liked to say. With her in such a state of confusion, the gettin' didn't get any better than this.

Al hopped again, but the movement drew Ann-Marie's attention. She stared at him, caught in a war between resolve and defeat. He stared back at her the way he stared at plebes in his academy days. "Give it up." He was quiet, menacing, in control. He was also out of ideas. "You know you can't do it."

"I will!" Ann-Marie's voice was frantic. "I am *not* a weakling!" She was looking at Al, but didn't seem to see him.

"Then why are you shaking like that?" he demanded. "You're terrified."

"*No!*" She turned the gun on Sam. "You don't scare me. You're at *my* mercy. I've devoted my life to reaching this point. Now you'll suffer, like you made me suffer."

"But I don't even—"

"Shut up, Sam!" Al yelled. The physicist's voice was the *last* thing she needed to hear. The sound of it would drive her to pull the trigger, internal debate or no. But why such hesitancy on her part? Why such terror? She said she had devoted her life to finding him; her hate was almost palpable. Yet the kill was a struggle. The pieces of the ugly puzzle came together in a flash. All Al needed to do was cast some light on the picture so she could see it.

"Well, here you are," he shot at Ann-Marie. "So why don't you end it? Why don't you have the guts to pull the trigger?" Al hobbled forward, not caring if she realized it. "I'll tell you why. You've centered everything on finding him, on putting an end to him. But once you do that, what reason do *you* have for living? If you've based your entire existence on hating him, what will you do when that net is pulled out from under you? What will you do if you don't have *him* to blame it on?"

"I'm not a victim!"

"If you don't have a scapegoat for your hate?" He had reached the halfway mark.

"I am not *weak*!"

"If you have nothing left to hide behind?"

"I am *not* a coward!"

"When you have no one left to hate but yourself?"

Ann-Marie screamed, the piercing sound echoing through the confines of the office. "*Shut up, shut up, shut up, shut up, shut up!*" She squeezed her hands to her ears, the pleading in her voice so tortured that Al felt sorrow for her. What was going on in there?

Tortured or not, his chance had come. The gun was aimed at the ceiling and her eyes were shut. He got almost within arm's reach of her before the cries died out and she snapped the gun back down.

"Right or wrong, it doesn't matter anymore," she said shakily, her voice small. "It's the only way to make them stop." Her tear-filled eyes met Al's. "You see that, don't you? It can't go on like this. I . . . I *can't*. Forgive me, Evangelene." She aimed the gun at Sam's head.

"No!" Al yelled. But before he had a chance to react, the door opened, hitting Ann-Marie from behind. She pitched forward, and he wrapped her in his arms.

A flurry of activity accompanied his tumble to the floor. He saw Sam's foot kick the gun out of Ann-Marie's hand, heard LoNigro yell "Dana!" and heard Sammy-Jo call his name over the familiar hoots of the hand link.

"I don't mean to bother you and Sam, Sibby," a familiar voice said testily as Al got the cuffs from the back of his belt, "but the phone in the lobby of Stratton was broken and Rex keeps beeping me. I just *know* it has something to do with that mud queen nonsense we saw on television. Can I use your phone?"

Al locked the cuffs around Ann-Marie's wrists without much struggle. She wept quietly into the carpet, her fight gone. He didn't know if her compliance signaled acceptance of what he had said or sprang from an entirely different source, but broken sobs were all she had left to hurl. In a way they hurt Al more than the bullet had.

He rolled off her back and sat up, turning his head toward

the door. The woman in green from the banquet hall stared at him in shock, voicing the question ricocheting around his brain. "What the hell is going on?"

He matched her dumbfounded stare until he felt a tugging at his arm. "Sam! You're alive!" His friend looked down at him, holding the gun and a large knife. For the first time, Al noticed Sam's shredded shirt, the spots of blood on the left side of his chest. He looked at the remains of the mason jar lying in a puddle of chemicals. The realization made him shiver.

Sam pulled him up. "Thanks to you," Sam glanced at the name tag, "Mr. Morgan. She dropped the knife when you called my name. I kicked it under the chair." He waved the weapons distastefully. "What should I do with these?"

Al pulled the walkie-talkie from his belt and waved the antenna toward the desk. "Dispatch, this is Morgan. I need an ambulance and every available officer at Eastman Labs, second floor. I left a light on for them." He clicked the unit off and hoisted himself onto the desk as Sam laid the weapons down.

"The time line is restored, Admiral!" The stability of Samantha's image had already told him as much. "Preparing for retrieval."

"Wait! What's going on?"

"My wife's impeccable sense of timing, I'm afraid," LoNigro said. He laughed, and gave the woman a squeeze. "Dana has a knack for finding trouble. I tried to insist that she keep her maiden name when we got married, so as not to ruin the family reputation. But she would have none of it."

The woman stood beside Al and fingered the gun Sam had dropped on the desk. As he stared at her, flashes of memory drove his jaw further toward his chest.

Sam handed Dana the gun, showing her how to use it and warning her to keep it pointed at him.

Al studied the hand link, relating what he saw crawling across the tiny screen. "She testifies in your behalf and keeps you from getting fifteen to twenty. Then she goes back to school . . . law school. Then in '76, she passes her bar and becomes senior... partner in the firm of Elroy and Elroy. And then it becomes Elroy, Elroy, and Lo-Ni-gro." He stared at the hand link to make sure he was reading it right. *"Sam, you're never gonna believe this...."* Sebastian LoNigro walked through the front door at that moment, laden with groceries. *"She marries...."*

"Professor LoNigro!" Sam exclaimed.

Al looked at him in askance. *"How did you know?"*

"Dana *Barrenger*?" She was older, streaks of gray peppering her blond hair, wrinkles where there hadn't been any before, but she was the same person Sam had saved during a Leap two years ago.

"Not for a long time," Dana replied absently. She was back at her husband's side, studying his wound. "Sibby, are you okay? Who *is* this woman?"

LoNigro shrugged, scratching his beard. "Ask Sam."

"Well, *I* don't know either," Sam exclaimed, looking at Al. "Mr. Morgan?" Sam repeated himself twice before getting Al's attention.

"Huh?" Al's mind was just absorbing the words Dana had spoken when she walked into the room. No, it *couldn't* be. "I don't *believe* this," he muttered, focusing on the woman. He was almost afraid to ask. "Did you mention someone named Rex before? Something about the mud queens?"

"Yes," the woman replied testily. "You've probably seen it on TV. The evil councilman Rex Barrenger is my big brother. He wants me to go to court with him."

Al glanced at Sammy-Jo. "What... what are the odds?"

The hand link dangled at her side. "A million to one. At *least*. I'd ask Ziggy to confirm it, but I'm afraid of locking her systems up again." The colored box squealed, and she raised it. "Retrieval program loaded. Hold on, Admiral. Time to take a ride."

Al's mind immediately sprang to a new problem. "Wait! Give me just a minute more!" Sam stood next to him, and Al grabbed his arm.

"There's no time left," Samantha said with a hint of sorrow. "The longer you stay, the less your chances of getting

210

home successfully. I won't lose *both* of you." She punched at colored buttons and nodded. "Affirmative, Gooshie. *Fire!*"

Al felt a tingle and clutched Sam's arm harder, pulling him into an embrace. All he needed was a little time to think. Was there a way for him to somehow turn this into a total victory, to change history again and prevent Sam from ever getting lost in time?

He knew the answer, of course. He hated it. But he checked his urge to warn Sam about rushing into the Accelerator Chamber in seven years' time—it would do more damage than Ann-Marie had ever been capable of. He settled for the hug instead.

"Looks like the homecoming will have to wait until another time, buddy," he said, not caring that Sam didn't understand what he meant. "But at the party *I* throw for you, the only thing you'll have to duck is confetti. It'll be a real kick in the butt."

Al held the embrace until the blue-white flash rendered him senseless.

CHAPTER TWENTY-ONE

Feeling started to return, and for a moment Al wondered if the retrieval had been successful. He still had someone cradled in his arms, but he was surrounded by mellow blueness. A familiar white dais came into view. The Waiting Room! He was home! But where was the slow music coming from?

He pulled back and found himself face to face with Sam. *Sam? What the hell is going on here?*

His friend had his eyes shut tight. He removed a hand from the small of Al's back to rub them. "What was that flash?" he asked. When he opened his eyes again, he recoiled in shock. "Who are you?"

He pushed Al away just as the former Leaper regained full feeling. Al's knee screamed at the weight it supported, and he tumbled to the floor with a grunt. Blood spread over the thigh of his Fermi suit "Ziggy!"

"Emergency medical team on the way, Admiral," the computer replied. "A pity I had to stop the music." It sighed. "The observation was becoming quite interesting."

"Where did you come from?" Sam asked again, his voice high. "What happened to Charlie?"

"Charlie?"

"The guy who was here a second ago. You were here first, then he was here, and now it's you again."

Al had never seen Sam pout before. "Charlie's gone. For good." He instinctively reached for his shirt pocket before realizing he didn't have one. "Ziggy, tell that crew to bring me a cigar!"

"This is one crazy dream," Sam said and retreated to the bunk. His falsetto voice grew petulant. "First an old guy, then a *real* man, now back to an old guy. What do you think that Freed guy would say?"

"Freud," Al corrected. "It's Freud. And I have no idea what he would say, but around *these* parts they call me Admiral Calavicci, not 'old guy.'" He shook his head. "I guess they call *you* Sarah?"

"Candy, if you like. Either or."

"Geez, it's a wonder Sam got through even one day of being you without anyone accusing him of being a pod person."

"What? I don't get it."

"I wouldn't worry," Al said testily. "I get the feeling that's something you're used to by now."

The silver-gray door slid up, and the medics burst in with a crash cart and gurney, Verbeena leading the pack. "Al, is that really you?"

"Depends on who you talk to. Can I get some help here?"

"*Ooohh*!" Candy squealed, "just like on *St. Elsewhere*! This dream is getting *good* again. Where's that hunky Dr. Axlerod?"

The entire medical team paused and turned toward the bunk. "You mean the television show?" Verbeena asked. "Wasn't that the heavyset guy with curly hair and glasses?"

"Yeah." Sarah lent Sam's eyes some of the dreaminess she put in his voice. "Now *he* was a real man."

The crew stood in silent wonder for a beat more before turning its attention back to Al. The medics hoisted him onto the gurney and wheeled him into the corridor. "What the hell was *that*?"

Verbeena was cutting the bloody area of the suit with bandage scissors. "Hmm? Oh, that's our current Visitor, Sarah Bullock."

"You *know* that's not what I'm talking about," Al groused. "Ouch! Be careful with that thing! I mean the mambo we were doing."

"Oh, *that*." Verbeena kept her head lowered, but Al could hear her laughter. "Well, when you Leaped out, we brought Charlie Morgan into the Waiting Room. He seemed calm enough, and we thought it would help convince both Sarah and Charlie that their Leaps were nothing more than a dream. One would voice the idea, reinforcing the other's suspicions. Simple, see?"

"Verbeena. . . ."

"I'm getting to it. They introduced themselves, but something odd happened when they shook hands. We didn't see any difference in either of them, but apparently they were able to see each other. It was as if the auras of Dr. Beckett and yourself canceled each other out when they met." Beeks shook her head. "Odd, that."

"Agreed," Al assented. "But not as odd as the waltz I Leapt into. Out with it."

"Well, these suits don't hide much"—Verbeena yanked on some of the stretchy fabric she had cut away—"and I take it Sarah is a girl of rather . . . generous . . . proportions. And Morgan, is he a little stocky?"

"Said tactfully and truthfully," Al replied. "Both times."

"Well, Morgan's spare tire showed Sarah that he had it where it counts so far as she's concerned. So one thing led to another. . . ."

"Don't tell me," Al shot. "They plunked a quarter into the old imaginary dream jukebox and started two-stepping their way into each other's hearts. Whose idea was the music?"

"A staff member thought it would be interesting to conduct an experiment on the effects that can be achieved when subjects with preconceived notions are introduced to new visual stimuli. . . ."

"Can the psychobabble, Beeks. I never figured you to shoot such dirty pool."

Verbeena looked up, her face the definition of innocence. "Admiral, surely you're not suggesting it was *my* idea?" She turned back to the injury on his leg and studied it with interest. "Truth be told, I found the whole thing rather unsettling. I mean, it was funny at first. But the image of you and Sam getting. . . ." She shuddered. "It was quite a spectacle. Or, to put it in terms I think Sarah would appreciate, the whole thing just weirded me out."

"As much as it did me?" Al shook his head. "Not by half, honey. Not by half. Come on, 'fess up. Who?"

"I doubt you'd believe me if I told you."

Al's tone was getting rough around the edges. "Ensign," he snapped to one of the medical crew, "whose idea was it?"

"I . . . I don't know, sir," the young woman replied.

Al looked at the other medics, but all of them suddenly seemed very interested in the walls. "If I have to start giving orders. . . ."

"Oh, knock it off," Verbeena said, "or I'll order *you* to be put on bed rest and a Jell-O diet. Just believe me when I tell you that you wouldn't believe me if I told you."

"What the hell did you just say?" He was fed up. "No more B.S. Tell me."

"Gooshie," Verbeena said. "It was Gooshie's idea."

"Gooshie?" Al pondered the name for a while before the truth sank in. "Gooshie? He actually made a *joke*? He made a joke!" Al laughed, all tension gone. "He made a joke, and a damn good one!"

Verbeena laughed with him. "And I don't think he's finished. I heard him talking to Ziggy about using a photo of the whole thing to design some Christmas cards."

Al laughed so hard he almost fell off the gurney.

Tina was waiting for them at the entrance to the medical bay and brought her hands to her mouth with a squeal when they rolled into view. "Al, honey! You're home! Oh, my *God*! What happened to your leg?" She ran to him and threw her arms around him. Al could barely breathe with the barrage of kisses she plastered on his face, interspersed with "Poor dear! Poor baby!" Then she planted him one full on the lips. She held it for a long time. He didn't resist.

"It's fine," Al said when he was finally able to take a breath. "I've taken much worse. Will you let them roll me in?"

Tina gave one last squeeze and backed out of the way, knocking a medic aside and taking a position at the head of the gurney. "Make way! Coming through! Medical emergency!"

"Thank you, Tina," Beeks said. "I think we have it under control now."

Tina moved to an out-of-the-way corner but didn't leave.

"What kind of damage are we talking about, Verbeena? Will I ever be able to slow dance again?"

Verbeena studied the wound, mystified. "This is the damnedest thing I've ever seen. You have all the trauma of a bullet wound, but there's no bullet. There's not even a powder burn. What happened?"

"Well, I sprained it pretty good at first. Then Ann-Marie shot me." Tina gasped. Al held up an arm, signaling her to stay put. "I figured I'd be lucky to get away with a permanent limp."

"There's no swelling indicative of a sprained knee," Verbeena said. "In fact, I didn't even think to check the joint until you mentioned it just now."

"What does that mean?"

"I have no idea." Verbeena shook her head. "If we're right in assuming that it's the entire body that Leaps, which I think Sarah and Charlie just confirmed for us, then the bullet wound is consistent with that. But since your knee is fine, I'd say it's not a hard-and-fast rule."

"It would *have* to be one or the other." Al insisted. "Wouldn't it?"

"That's a discussion best left for Samantha, or Sam when you see him next," Verbeena said, pushing the air bubbles out of a filled syringe. "But I, personally, have seen evidence to support both. Remember Billie Jean Crockett, the pregnant girl we had here a while back? First her baby Leaped with her, but then it was gone. That implies it had a womb to go back to."

"But what about Ronald Miller, that Marine captain Sam

Leaped into who lost his legs? Sam was still able to stand up, regardless."

Verbeena shrugged. "As I said, it's a conversation that's out of my depth. I just fix 'em if they need fixing." She held up the syringe. "This is a local. It's going to sting a little." She shoved the needle into Al's thigh.

Sting? He gasped. It felt like.... The pain was suddenly gone, not just the bite of the needle, but the relentless throbbing and burning as well. "*That's* the stuff," Al said. "Now fix me up so I can get out of here." A thought struck him. "Does that mean Morgan is Leaping back to a sprained leg and no bullet?"

Verbeena shrugged again. "Got me. Obviously, you couldn't bring the bullet with you when you left...."

"But it never entered *his* body. What will happen to it, I wonder?"

"Maybe his Leap in will displace it from our continuum," Tina said. "Maybe there's, like, a whole 'nother dimension where it will float free for eternity." Her eyes lit with inspiration. "It's probably the same place Dr. Beckett goes when there's a lot of time between Leaps." She smiled. "Maybe, like, Dr. Beckett will have a friend waiting for him every time he Leaps now...."

"A pet *bullet*?" Al was incredulous.

"That's enough, you two!" Verbeena snapped. "You're going to give me nightmares of what would happen if Sam Leaped into someone with a pacemaker, or a steel plate in their head." She ripped open a suture package. "Let's just take it for granted that higher forces are watching this whole thing and exerting some control."

"I'd say we saw that theory in action just now."

Al turned and saw Sammy-Jo standing in the doorway. She was grinning from ear to ear, practically beaming. "Glad to see you, too," he said.

"It worked," Samantha cried. "The damned program *worked*!" She laughed in triumph. "I'll get him home yet!"

"What did you mean before when you said you saw the theory in action?" Tina asked.

Samantha shook her head. "It was the strangest thing. In a way, Dr. Beckett solved the problem himself."

Al took over when he saw the questioning looks on the surrounding faces. "There's no time to explain it all right now," he said. "But I think things are more connected than we think. Sam's current Leap into Sarah Bullock set events in motion that helped us stop Ann-Marie. And even *that* wouldn't have been possible if not for one of his Leaps from a few years back." He shook his head. "When I think about what *really* saved us, what I did seems more like moral support than anything else."

"That's not entirely true, Al," Samantha said. "There's no telling how much time you bought, what would have happened had you not been there to distract Ann-Marie. And Professor LoNigro's wife would never have had cause to go to his office if you didn't break the phone." Samantha grew pensive. "I think we're looking at this whole thing from the wrong perspective."

"How so?" Al asked.

"Well, it's actually kind of funny if you think about it, but we were all so concerned about the bit of foreknowledge Ann-Marie managed to get out of her Leap experience that none of us even considered the possibility that there are higher forces that already know how events will play out." Samantha extended her arms in an all-encompassing arc. "Higher forces that are controlling matters, working to make sure things happen the right way. There's even a scientific line of thought that postulates something along the same lines. It says a paradox can never occur, that time *can't* negate itself. Any attempt will become its own undoing."

"It doesn't sound like a matter of science to me, Samantha," Verbeena said, taking the last stitch. "It sounds like a matter of faith."

"My, but that's a *colloquial* sentiment, Dr. Beeks," Ziggy said with a hint of disdain. "And as much as I hate to break up this homecoming celebration, I must inform you that Dr. Beckett requires a little more than *faith* to complete his mission. He is in need of assistance."

"Oh, *geez*, Sam!" Al hopped off the table, and his leg gave out. Pain wasn't the problem any longer. It was that he could feel absolutely nothing.

"Al, get back up there this instant!" Verbeena barked, helping him up.

"No can do, Doc," he said, testing the joint. His knee looked to be bending fine. It was just a matter of getting used to walking with a sensory hole in one leg. Piece of cake. "Fun's over. I gotta get back to work." He stumbled to the door, where Tina waited with supporting arms. He nodded at Sammy-Jo. "You come, too."

The three were in the hall, Al setting as fast a pace as he could muster. "What eventually happens to Ann-Marie?"

"She gets tried for attempted murder, but is found not guilty by reason of insanity," Samantha replied. "But Ziggy said she *does* get locked up again for smuggling. She does the time in a mental facility. As a matter of fact, she got out last year."

Al continued to walk in silence, his mood worsening with each step. By the time they reached his quarters, he was able to walk unassisted. He disappeared into the bedroom and quickly began dressing. The source of his ire was simple, and he voiced it through the half-opened door. "What did we do to that poor woman?"

"Who?" Sammy-Jo called from the outer room.

"Ann-Marie."

"Don't you have that backward?"

"I wish I did." Al sighed, buttoning up a maroon shirt with canary sleeves and collar. He felt crummy, despite the bright colors. "I wish I did. But what that woman did, she did as a *direct* result of the path she found herself on after Sam Leaped out. We drove her to it. There's no getting around that." He grabbed a few cigars from the humidor on his desk and popped one in his mouth, depositing the rest into his shirt pocket. He removed the stogie from his lips and started to unwrap it, but stopped in midpeel when he saw Sammy-Jo's expression. He stood in the doorway, meeting her indignant stare with a serene one. "You're going to deny it?"

"You can't put it in such simple terms, and you know it," Samantha said. "Verbeena gave me the psychological profile too, remember? That woman was unstable from the start, with no respect for anyone or anything, and no apparent rationale for the decisions she made. The very fact that she was institutionalized suggests she has a chemical imbalance, at least. She may even have an organic brain disorder."

"I'd say her rationale was *very* apparent," Al replied testily. "Your neural link was almost nonexistent. If you were getting through any better, you'd already know that the psych profile doesn't wash our hands entirely clean, even if it *is* right. You didn't hear the anguish in her voice. The pleading. I chided her for not taking responsibility for what she's done in life. How am *I* any better if I don't do the same?"

"But it's *not* the same," Samantha replied with impatience. "You and Dr. Beckett were only doing what you could, what you knew to be right at the time. That's not good enough anymore? Suddenly you're responsible for any hiccup that might occur further down the line as a result?"

Al rubbed his forehead. "Is it just me, or does this conversation sound really familiar?"

"Just give me a pool of banana-fudge chunk." Samantha's tone softened. "Just because the shoe is on the other foot, it doesn't make the advice any less valid. In a way, you're even better off. I should remind you that if you *didn't* do what you did the first time around, Dr. Beckett would most likely be getting used to his new life as Ann-Marie Renerie. You had no choice. There's nothing that can be gained by second-guessing every move you make. If you do, you'll always be in your own way. You'll never move forward."

"You're right"—Al nodded—"to an extent. But I don't have to like it, and I don't think I'll ever be able to justify the part I played in the whole thing."

"Nobody is asking you to."

Al lit the cigar and rubbed his hands together. He would let the questions wait until he had time to dwell on them properly. "Let's see to Sam."

The Control Room was awash in its usual cascading color patterns. Gooshie, at his place behind the console, offered a hand link. "Good to see you back so soon, Admiral."

"Good to be back." Al took the hand link, shook it once for good measure. It squealed suitably. "And don't forget to send me a Christmas card this year." He started up the ramp to the Imaging Chamber but stopped when he heard someone clearing her throat. "No hello for me, huh?"

Al turned slowly. Donna stood behind him, hands on hips, foot tapping. "Donna!" He ran down the ramp and threw his arms around her. "Boy, is it good to see you!"

"All right, already!" Donna laughed. "You know I'm not really mad."

Al held on anyway, making sure she was real. Not only had he not missed her for the last week; he hadn't even known he was *supposed* to have missed her. How close they had come.... "I know it," he said. "I'm just glad to see you. Forgive me if I'm being overly mushy."

"Maybe *she* can, but I don't know if *I'll* be able to." Tina struck the same pose Donna had been in a moment ago, only Al could tell she wasn't fooling. "All *this* for her, but all you give *me* is one lousy kiss!" She turned and stamped to the console, heels clicking like machine guns on the concrete floor.

"Tina, honey!" Al pleaded, taking a step after her. "I was wounded. Laid up!" She turned and he stretched his arms toward her. "You understand, don't you?"

"I understand perfectly, little Mister Welcome Wagon." She rounded the console and sidled up next to Gooshie. Al saw only one of her hands resting on the blinking display. What was she doing with the other?

Gooshie's face suddenly went as red as his hair. "And when you get back from playing with Sam, don't come looking for me." She stared longingly at Gooshie. The head programmer swallowed and ran a finger under his collar. "I have a feeling I'm gonna be busy. You, like, *understand*, don't you?"

"Perfectly," Al mumbled, shaking his head. "Perfectly." He reached the top of the ramp and punched the security code into the control panel beside the door. The steel barrier slid up and he walked into the blue haze, aiming for the silver disk at the room's center.

He waited for the door to come back down before he let out a long, low laugh. He understood that things were back to normal, or what *passed* for normal around Project Quantum Leap. All except.... "Ziggy?"

"Yes, Admiral?"

"How long do you think it will be before the committee brings me up on charges for going against its orders?"

"Orders, Admiral?"

"They told me not to Leap. Expressly forbade it, to be more accurate. Remember?"

"Well, yes, of course *I* remember. What a *silly* question!" The computer let a precise amount of silence pass to show it was insulted before it continued. "*You* probably remember as well. *They*, however, never made those orders in this time line."

"Huh?"

"You averted the threat posed by Ann-Marie Renerie, correct?"

"Yeah."

"In doing so, you changed the original history. To those outside the Project, and even to those in it who had no direct dealings with the situation, none of it ever happened."

"So you're telling me. . . ."

"Why do humans always need a picture painted for them?" Ziggy asked testily. "It's at times like this that I miss Dr. Beckett the most." The computer gave the equivalent of a sigh and continued in a lecturing tone that would insult most kindergartners. "Think of time as a river. The entire week that has just passed, as you remember it, is a branch of that river that breaks off and dries up. The river itself, however, has made minute adjustments to its original course and roars on. It seems no different except to those who paddled into the dead end and had to find a way back."

Al understood perfectly what she meant, of course, even without the river analogy. He just enjoyed baiting her. It was a good way to keep sane, a diversion that kept him from concentrating too hard on the circles within circles, the confusing ripples of time travel.

He stepped onto the silver disk and made the hand link sing. "Begin program, Gooshie."

CHAPTER TWENTY-TWO

The police station loomed ahead, the barriers set up in front of it seeming to bulge with the surge of the crowd. Sam, Kyle, and Tawny had decided to walk the five blocks from town square; there was no way to get a car through the mess.

A second, human barrier of broadcast reporters lined the rear edge of the milling mass, their microphones raised at regular intervals like readied rifles as they recorded their stand-ups and delivered live reports.

His job was finished, so far as Sam could tell. Barrenger was exposed, Kat was on her way to becoming mayor, and Kyle and Tawny had not snapped at each other once. In fact, they were getting along famously, recounting the events of the evening with laughter and a mutual sense of accomplishment.

Sam joined in the conversation halfheartedly, worry gnawing at his elation. It was all over, even the shouting, and still there was no sign of the Observer and no Leap feeling.

"She should be around here somewhere," Kyle said as they passed behind the ranks of reporters. "There she is! Oh, and look, the little nimrod is there as well." Kyle rushed ahead.

Sam looked at Tawny, eyebrow raised. "Our mysterious source?"

"Let's find out."

Sam grabbed her arm. "Hang on a minute." He quickly took his hand away and raised it in surrender. "I guess I have a beating coming my way."

Tawny's expression was questioning, but she put a hand to her mouth in surprise after a moment. "Oh, *that*. I'm sorry I threatened you. I was just pissed off, and you caught me at the exact wrong time."

"It's okay. We've all been there. I just wanted to see if anything has changed between the two of you. I mean, it looks like you're getting along better."

"We are." Tawny smiled and rubbed Sam's forearm. "Thanks for trying to help. We haven't gotten along this well since before Mom died. Hell, I don't remember *ever* getting along this well."

"Sometimes it takes a tragedy for us to realize how much we really mean to each other," Sam said, "or to realize that family members are probably the most important people in our lives. Even if they *do* have a tendency to get on our nerves."

"That's what keeps it interesting, I guess."

"Are you going to go see your dad?"

"Kyle's told me how much he's changed." Tawny sighed. "I don't think he'll ever be the same man I grew up with, but I have to try."

Sam felt a twinge of worry. "What if it's still not so easy? You won't leave again, will you?"

Tawny shook her head. "It won't really matter, to tell the truth. I've decided that I'm not moving back home, no matter what happens. It would be like taking a step backward, for all of us. Who knows? If I'm constantly there, Dad may decide he no longer needs his newfound independence. That wouldn't be healthy." She paused. "But I'll never turn my back on them again, if that's what's bothering you. I am who I am. If they can't handle it, that's *their* problem. They'll always know where to reach me."

Sam gave a tentative nod. "I see your point." It wasn't all smiles and rainbows, he thought, but at least the lines of communication were open. There wasn't anything more he could do. "I guess that's all *any* person can ask from another, family or not."

"Well, I hope that Kyle sees it that way, too. He's a little stubborn, and I get the feeling he won't take it as calmly." Tawny rolled her eyes. "When I get the nerve to tell him, that is."

"I don't think he'll give you much of a problem anymore," Sam said. "He's seen that you can take care of yourself. He'll never be crazy about the mud wrestling, make no mistake about that, but he knows that he doesn't have to worry about you. I think that's all he was looking for in the first place."

"Girls!" Kyle waited a few feet ahead, pacing impatiently. "Let's go!"

Sam and Tawny trotted forward, following Kyle's lead. He loped toward a man and woman who were arguing. Sam assumed they were reporters, since the woman held a microphone. The man paced in front of a camera that had been set up on a tripod.

"You can't do this!" Sam heard the man exclaim once they were in earshot.

"Look around you, Jerry!" The woman yelled, waving her microphone at the crowd. "The story's broken. You can't hide it any longer!"

Kyle halted a few feet away. "Is there a problem here?"

Jerry turned on him. "*You*. I should have known you'd be mixed up in this!"

"Looks like you got me, Jerry boy. It's a wonder you're not working in a bigger market, what with that uncanny deductive reasoning you display."

"Where do you get off, coming in here covering *my* beat?"

"From what I understand, you've been busy covering other things," Kyle shot back. "Or should I say covering them *up*, Browne? Nancy asked for my help because she knew that if you got wind of what she'd found, she'd be out of a job."

"What's going on?" Tawny asked. "Is this the source you kept mentioning?"

"Let's make some introductions, shall we?" Kyle extended a hand toward the woman. "This is Nancy Dotorovic, and she *is* my source. Her sparring partner here is Jerry Browne." Kyle slapped his forehead. "I'm sorry, Jerry," he said. "How insensitive of me. Browne is just your on-air name, isn't it? Please permit me to take that back, ladies. This is Jerry Barrenger."

"Jerry *Barrenger*?" Sam exclaimed. "Is he...."

"Old Rex is his second cousin." Kyle scratched his chin questioningly. "Or is that first cousin once removed? I never was very good at family relationships...." He winked at his sister. "But whichever you call it, I'm right, aren't I, Jerry? You were pulling double duty as cousin Rex's personal censor, making sure the story didn't leak."

"How dare you insult me like that!" Browne sputtered. "My reputation as a reporter in this town is spotless! Ask any—"

"Then why did Nancy feel she couldn't turn to her partner for help with the biggest story this town has seen in years? Why did she feel she couldn't even share it with anyone in the newsroom? And why are you still trying to stop her, even now? Admit it! You're in his pocket, and if you knew about *any* of this earlier you would have gone running."

"Where?" Browne was indignant. "Even if what you say is true, how could Councilman Barrenger possibly have done anything to prevent the story from getting out? He doesn't own the television station."

Kyle shrugged. "He's a powerful man. *Was* a powerful man, anyway. I'm sure he would have been able to find a friend or two in the proper position to get Nancy a one-way ticket out."

"I don't know where you're getting all this from"— Browne shook his head, a resigned grin on his face—"and you obviously refuse to listen to the truth. But you're dead wrong."

"I guess we'll just wait and see," Kyle said. "The truth will come out eventually." He stared hard at the man. "We both know it. And even if Barrenger doesn't finger you directly, I'd say your credibility is shot." Kyle's stare grew deadly. "I'll see to it myself, if necessary. That ought to wipe that stupid grin off your face."

Browne did his best to match the stare, but soon lowered his gaze and stalked off. He turned once, mouth open, finger raised, but if he had anything left to say, he thought better of it. He disappeared into the crowd.

"Good riddance," Kyle mumbled. "Teach you to mess with *mia familia*."

"Thanks, bud," Nancy said warmly, giving Kyle a punch in the shoulder. "Now get your butt behind that camera so I can shoot this stand-up and get inside."

Sam stood aside with Tawny, letting Nancy do her work. His mind chewed on the latest revelation. Of all the people who could have been Kyle's source, he never in a million years would have guessed that it was another reporter. That wasn't even something Ziggy would have suspected.

But was what the parallel-hybrid computer suspected or didn't suspect really all that important? He felt that he had done his job rather well, Ziggy's guidance notwithstanding. Hell, it was a Leap that under normal circumstances he would have considered easy. All he really did over the course of the last week was make a few suggestions, lend some support where he thought it was needed. Things had rolled with their own momentum from there. The only thing that had been missing was the creature comforts.

Another thing he realized was missing, finally, was the feeling of dread that had been looming in his subconscious. As glad as he was to be rid of it, however, he tried to recall the sensation, hoping that he would remember it in future Leaps. Maybe it was really a gift bestowed upon him by whatever forces were directing his movements through time—a sort of mental barometer that he could count on if he ever found himself flying solo again.

"You guys ready?" Kyle asked, shouldering the camera and tape case. "Nancy says she wants to go inside."

"Yeah," Tawny said. "Her and everybody else. How are we going to get through that?"

Sam's eyes followed her extended finger to the entryway. The sawhorse barricades were doing an excellent job of keeping the crowd back, townspeople and reporters alike. Apparently no one was getting in.

"It's all been arranged," Nancy said as she looped up the microphone cord. "We won't have any problems."

Sam doubted it, but followed the others to the front of the mob. Nancy spoke briefly and quietly with one of the patrolling officers, and the man let them pass.

"What did I tell you?" she said as they entered the lobby. "Piece of cake." She had an equally short conversation with the desk officer before they were escorted to their final destination.

Kat and Richard sat on chairs lined against the wall of the short corridor, Allison between them. "*There* you are," Kat said as they approached. "I was getting worried that you bailed out on me."

"Why?" Sam asked. "We're not taking the heat, we're giving it. Where's Barrenger?"

Kat tilted her head toward the closed door on the other side of the corridor. "In there with McClough and the police chief, probably trying to save what's left of his skin. I was beginning to think I'd have to do the same without my co-conspirators."

"Kathy"—Kyle stretched the name, holding his arms out—"have I let you down yet?"

"Save it," Kat snapped good-naturedly. "I wasn't talking to you." She approached Nancy. "You, my dear, are brilliant. Simply brilliant. Allison told me you were the one who orchestrated this whole thing. We were having a rather enlightening conversation."

"Don't let her fool you into giving me all the credit," Nancy replied with a shake of her head. "This was Allison's baby from the start. I just used what she gave me."

"What are you two talking about?" Sam asked.

"Let's just say that Barrenger isn't the only one in this town with a well-placed cousin." Nancy approached the girl sitting next to Richard. "Mine is just *much* smarter. My hat is off to you, kiddo. But please don't decide to go into journalism. I don't need the competition."

"Me either," Kyle said.

Allison grinned sheepishly. "You know it's not like that."

"Like what?" Sam asked. "What in the world is going on?"

"Go ahead," Nancy said. "Tell her."

"It's all pretty simple, really," Allison said to Sam. "I was at my lousy job in the hall of records. Lois asked me to organize the files and I found that bid—you know, the trash bid written in pencil?"

Sam nodded.

"I found it in the back of a drawer where it didn't belong," Allison continued. "When I asked Lois what to do with it, she practically ripped it out of my hands, put it back where I found it, and told me to forget about it."

"Don't you just love it when you meet a real people person?" Kyle asked. "I guess she thought one command would suffice, since to her, young people are the champions of laziness and stupidity."

Allison rolled her eyes. "You don't have to tell *me*. But I remembered my civics teacher mentioning that those bids were supposed to be written in pen for them to be legal. I knew something had to be up, so I told Nancy about it."

"We did a little more digging, and you know the rest," Nancy said.

"Not quite." Sam raised an index finger. "Why did I have to tell you 'Nevenka' sent me when I picked up the folder?"

Allison turned to Nancy, a knowing smile on her lips. "Cuz?"

"Browne isn't the only one with an on-air name," Nancy said, smirking at her cousin. "Nancy's just a nickname. My real name is Nevenka."

"Nevenka," Sam repeated. "No one would think to use *that*. It has kind of a catchy sound, like the name of a talk show. Much better than Nancy, if you want my opinion."

Nancy shook her head slowly, staring at Sam as if he had sprouted another ear. "You're woefully ignorant of the broadcast biz, aren't you?"

Sam shrugged, somewhat taken aback. No one had ever accused him of being woefully ignorant about *anything*.

"I'm already making their hair stand up with a last name like Dotorovic. If I went by Nevenka, any news director who saw my resume would begin to hemorrhage." Nancy shook

231

her head. "I appreciate the input, but I'd like to stay employed, thank you very much."

"Speaking of employment"—Kat broke in—"I have the feeling that as of tomorrow, I'll have a new job to worry about. The good part about that is that the club can stay open. Unfortunately, there will be no one to run it come January." She turned to Sam. "Unless, of course, *you'd* consider taking the job?"

"I. . . ."

"Don't do it, Sam," came a gravelly voice from behind. Sam felt a flood of relief and surprise. A flash of anger followed, and his instinct was to wheel around and face the source of the voice. It was all he could do to keep his feet planted.

". . . really need to go to the bathroom," he finished. "If you'll just excuse me for a second?" He practically ran around the corner.

"Haven't we already talked about this problem of yours, honey?" Kat called after him.

Sam burst into the men's room. A glance in the mirror reminded him he was in the wrong place, and he bolted back into the corridor, in search of the appropriate door.

The Observer stood calmly in the corner of the ladies' room, ash disappearing as he flicked it off the end of his cigar. "Get to the urinal and discover that you didn't have the right parts? You didn't have to meet me here. We were fine with the others."

"The *hell* we were," Sam shot back. "Where have you *been*? Is everything okay back at the Project?"

"It's fine," Al said, pacing toward his friend. Sam couldn't point to anything specific, but Al seemed more relieved than he himself was. "I'm sorry it took us so long to get to you, but I was called away. . . ."

"Why are you limping?"

". . . was called *away* to another alimony hearing," Al continued. "Only this time I won, and Sharon gave me a shot with her heel so the victory wouldn't be as sweet."

"Figures," Sam said. "Is Gooshie okay? Is the problem with Ziggy fixed? What was it?"

"Oh, *that*? Yeah, turned out to be nothing. Boy, you need

to learn to relax a little more. Gooshie just installed a circuit board backward or something. I would have been here a little sooner, but since we hadn't contacted you in so long, her majesty's proximity circuits were off. I wound up in the middle of the circus outside. Looks like you had a busy week." Al clenched the cigar in his teeth and retrieved data from the hand link. "A *very* busy week."

"To say the least," Sam said, taking a deep breath. "You don't know how *good* it is to see you again."

"I missed you, too. Especially when the lawyer asked for character witnesses." The Observer shook the hand link. Sam enjoyed hearing the tinkle. "It looks like you've pretty much wrapped things up."

"Yeah, that's what I figured. All I have to do is accept that job Kat is offering and—"

"You're *not* gonna take that job," Al cut him off. "Ziggy's not giving it good odds."

"What do you mean? It's the perfect opportunity for Sarah."

"No, it's a good opportunity for you, not her. Sarah'd never be able to handle it."

"Why not?"

"I had a chance to . . . talk to her earlier." Al spun the glowing tip of the cigar around his ear and whistled like a cuckoo clock.

"That bad, huh?"

"Let's just say that she's not the sharpest knife in the drawer," Al said. "I think she became a mud wrestler so she wouldn't have to worry about the complexities of tying her shoes before going to work in the morning."

"Al!" Sam chided. "Couldn't you be a little more kind?"

"That *is* kind, believe me."

"What's supposed to happen, then?" Sam's focus drifted again to Al's limping gait. A rather uncomfortable part of the evening sprang back to his mind. "Does Ziggy say anything about Sarah and Kyle?"

"What, you mean romantically?" Al cocked a thumb over his shoulder and chuckled. "Old beanpole back there doesn't have a chance."

"What, then?"

Al squinted at the minuscule screen. "Meet me back with the others." He pressed some buttons and popped out of sight.

When Sam reentered the corridor, the Observer was standing next to Tawny. "*She's* the one that should take the job, Sam. Ziggy's giving it ninety-five percent."

"How 'bout it?" Kat gave him an expectant stare. "Does my club have a new manager?"

"It'll have a new manager," Sam said with a smile, "but it's not going to be me, I'm afraid."

"Why *not*?" Kat sounded stunned. "With all you've done this past week, it's like you're a whole different person. After seeing you in action, I just *know* you'd be a natural."

"Maybe so," Sam replied, "but I'm afraid that's not where my interests lie. I do, however, know someone who would do just as good a job. Probably better, in fact."

"Who?" Tawny asked.

"You," Sam said.

Tawny's eyes grew wide. "*Me*? What ever gave you that idea?"

"Just call it a hunch," Sam said, tapping his chest. "It feels right to me."

"Considering how far I've gone on your hunches...." Kat turned to Tawny. "What do you say?"

Tawny's mouth hung open. "I...." She threw Kyle a questioning glance.

"Don't look at *me*," he said, hands raised in front of him. "You're the one who keeps telling me to butt out."

Tawny gave her brother a smile that Sam found reassuring, then turned back to Kat with a determined air. "I'll take it."

The hand link squealed with renewed vigor. "That did it, Sam," Al said. "Tawny manages the club for a few years and then opens a place of her own in Manhattan. No mud wrestlers." He sighed. "What a pity. But it'll do a booming business, become one of the hottest hangouts in New York City." Al paused. "*This* is interesting. Kyle will be in New York, too, working in television news. It looks like this mud queens story will do a lot for his career. And their father will get a nice little place of his own not far off, in White Plains."

The door to the police chief's office opened and Mc-

Clough stuck her head out. "Ms. Danson, we're ready for you." Barrenger stalked out with a police escort. For once he remained silent, but the look of contempt he cast on the group said plenty. "You, too, Ms. Wottawa. Have your parents arrived yet?"

"They're on their way," Allison said.

"I'll be acting on their behalf until they get here," Nancy said. "I'm her cousin."

"Well, knock me over with a feather, Ms. Dotorovic," the mayor said with a smile. "Who would have figured? I guess that means you'll have one heck of an exclusive interview. Let's get this finished. I have to get up early tomorrow so I can vote." The three women filed into the office and the door clicked shut behind them.

"Anyone hungry?" Kyle asked. "I could go for a little something. Any takers?"

"Count me in," Tawny said, sliding a hand into the crook of her brother's arm. "Coming, Candy?"

Kyle looked at Sam expectantly. Too much so. The Leaper quickly shook his head. "Uh, no. You go on ahead. I'm pooped. All I want to do is hit the hay."

"Your loss," Kyle shrugged. "Mr. Danson?"

"I appreciate the offer, but I'll just wait here until Kat comes out."

"Catch up with you guys later, then," Tawny said.

"I'd better take the time now to thank you for your undercover work," Kyle said. "I'm afraid of the penalty if I forget. I don't relish the prospect of another shot to the head."

"You don't have anything to fear," Sam replied, narrowing his eyes at Kyle, "*yet*. And you're welcome." Kyle looked slightly shaken as his sister pulled him down the corridor. Sam stared after him, tapping the skin under his eye. Kyle instinctively reached for the same spot on his own face before turning the corner.

"Good work, buddy," Al said. "He was barking up the wrong tree, anyway. Unless he suddenly develops a penchant for Twinkies...."

Sam raised an eyebrow at his friend, but a soft voice behind him stopped him from pursuing the question. "Sarah?"

"Yes?" Sam turned and saw Richard there, hand extended. He shook it, ladylike.

"I owe you some thanks as well," the older man said, "for a whole new future of interesting possibilities. I'll have to bone up on my skills as town's first man."

"Shouldn't be too much of a problem," Sam said. "I think you'll have to start another family photo album, though. I get the feeling Kat's picture is going to show up even more now than it used to."

"Oh, Kat will *love* that," Richard said mischievously. "Now go. Get rested while you can. In a few days you'll have a job to go back to."

"Good night," Sam said, retreating around the corner. "What happens to them, Al?"

"Kat wins tomorrow's election by a landslide," the Observer replied. "She goes on to serve another term before she and Dicky retire permanently. They stay right here in Wilson. As for that other reporter, she gets her own talk show based out of Chicago."

"Don't tell me," Sam said. "It's called *Nevenka*."

"Bingo," said Al. "How did you know?"

"Another hunch."

"Her cousin Allison becomes an author...." He punched the hand link, eyebrows raised in surprise. "In fact, her first book came out last year. 1998, I mean. And you'll never believe this—it's a sci-fi time-travel story."

"A very perceptive young lady," Sam said. "I could tell that the first time I spoke to her. What about the man whose empire she toppled?"

"Well, Barrenger does some time for fraud, but when he gets out, it's politics as usual. He successfully runs for a Republican Senate seat in 1994."

"Ah, the great game." Sam sighed as he walked out into the cool night. "You just can't help but love it." The crowd had thinned out considerably. The only people still hanging around were reporters, probably waiting for a chance to pounce on Kat when she exited. "So that should do it. Do I have to ask the question?"

"I know, I know. Why haven't you Leaped yet?" Al puttered with the hand link some more. "Here it is. It seems that Kat makes one final big move as manager of the club. Call it a grand exit. When she and Richard go there tomorrow, they'll find a representative from a production company waiting for them."

"Yeah, so?"

"The guy is gonna offer them some cash to film a special that will air on cable stations nationally. It's one of those pay-per-view things." Al batted at some glowing buttons. "*The Mud Queens Mud Wrestling Extravaganza*, they'll call it."

"What does that have to do with my Leap?"

"Ziggy says that a rep from a modeling agency will be watching. . . ."

"Oh, no," Sam said. The detail he had been so worried he was missing suddenly snuck up from behind and bit him on the tush. "No, no, *no*."

"Yes, yes, *yes*," the Observer shot back, grinning around his cigar.

"Don't say it, Al. Don't *say* it."

"Ah, is your bikini clean?"

Sam ducked through the ropes, grimacing as mud slithered up between his toes. Through the smoke and flashing neon he saw his opponent entering the far corner, wearing a very determined look on her face and not much else.

No matter, Sam decided. For once, he was going to use every advantage open to him—his superior strength, his martial arts skills, whatever else he could think of. He was going to *trounce* her.

Al had only said Sarah needed to be seen on camera. He didn't say for how long. If Sam got his way, it would be the shortest match of the night.

"Oh, cut the sour puss," Al said. He floated three inches above the middle of the ring, looking longingly at the other girl. "What I wouldn't give to be in your shoes right now! Metaphorically speaking, of course."

Sam waved him over, clutching a rope with the other hand so he wouldn't slip. Al floated leisurely to his friend, blowing kisses at Sam's opponent. "Tell me again why I'm doing this."

"Why?" Al moaned. "Why the hell would any sane man need a reason? Don't worry, Mr. Morals, it's for the greater good. Ziggy says it will propel Sarah into a successful modeling career. She'll never have to worry about winding up on the streets again. Is *that* enough to appease the Boy Scout in you?" Al stole another glance over his shoulder. "Personally, I don't see why you would need a convincer."

The bell dinged and Richard went into his announcer's spiel. "A-Ladies-and-a-gentlemen," his voice boomed over the music, "we have now entering the ring. . . ."

"This is *it*, right?" Sam asked, adjusting the strap of his bikini. "No more surprises. I don't have to go on to be ultimate mud queen or anything?"

"Unfortunately, no. One slip-trip'll do you."

"But why can't Sarah do this herself?" Sam asked, exasperated. "I mean, she'd probably be a lot better at it. What can I do that she can't?"

"You mean besides count to twenty without taking off your shoes?" Al's face sobered at Sam's warning look. "Ziggy says that in the original history, she slipped when she left the ring and broke her nose. As a result, the modeling offers never came her way. Just put one foot successfully in front of the other and you can wave good-bye."

Sam nodded, rolling his eyes skyward. "I expect some payback for this one, you hear me? The next Leap better be into a resort in Tahiti where all I have to do is prevent someone from getting too much sunburn."

"Don't worry, buddy," Al said seriously. "They're watching. They care."

The bell dinged three times, preventing Sam from responding. He ran into the ring as best he could with the slippery footing and met his oncoming opponent before she reached the halfway point.

He slid in low and swept her legs out from under her. Before she could even gasp, he had her shoulders pinned. She grabbed his head defiantly, driving his face deep into the ooze. He couldn't breathe, but he didn't care. He just had to hold on for the count of three.

The pressure on the back of his head finally ceased and he lifted his face, breathing deeply and swiping mud out of his nostrils. He then helped his opponent to her feet and hopped out of the ring.

"We have a winner," Richard announced, "Miss Candy *Apple*!" He held the microphone out for Sam to give a victory speech.

Sam gave him a smile, then walked through the bar in silence, ignoring the hoots and whistles. He paused briefly and smiled into the camera before he retreated to the safety of the hallway.

"That did it, Sam!" Al was suddenly beside him. "It was enough to get her seen by the right people, schnoz intact."

"Well, thank God for small fav—"

Kat burst through the door Sam had just closed behind him. She ran down the hall, passing right through Al and bumping Sam hard enough to make him stumble. A glob of mud ran down his forehead, stinging his eyes. He reached blindly for a wall and began feeling his way to the dressing room. Before he got there, he heard Kat open the side door and gasp.

"I *thought* that was your car I saw pull in," she said hurriedly. "Boy, did *you* pick the wrong night to visit, Governor. There are cameras all over the place!"

Al let out a surprised gasp of his own before the corridor was filled with gales of his throaty laughter.

Sam was still wiping mud out of his eyes when he Leaped.

EPILOGUE

Al walked slowly down the hall, wondering how to proceed. He had taken this trip once, sort of, and this time felt just as dangerous as the last. It was even worse, in a way. If there was anything he hated in all the world, it was state-run institutions. He wrinkled his nose at the oppressive smell of the place and fought back memories he had never been able to fully face.

He found his destination and held his breath, knocking softly on the half-opened door. "Ann-Marie?"

The woman sat in a chair, staring out a window at the grounds beyond. It was her second trip back to the hospital since she had been officially released eight months ago. Ziggy had not been able to locate any addresses for her in the times between hospital stays. Al didn't need the computer to tell him what it meant. Cast out with no structure, no guidance, Ann-Marie lived on the street until her illness overcame her and forced her to come back again.

He walked slowly into the room, noting the few hand-drawn pictures that hung on the walls. All were of the same smiling black woman. Al called to Ann-Marie softly once again.

The woman drew her focus away from the outside and slowly turned her head toward him. She was older than Al remembered, and she looked so tired. He knew it had nothing to do with lack of sleep.

Al sat slowly on the bed, not knowing how to begin. "Uh, I'm...." He waved the forms awkwardly. "These... these are sponsorship papers. They mean that I'm the one who's responsible for taking care of you from now on. You won't have to live on the streets anymore, and you'll always have money for your medication. It means you can start to put your life back together."

Ann-Marie furrowed her brows slowly, looking at him quizzically through her drug-induced haze. "Who are you?" she asked softly.

"I'm a friend," Al replied, taking her hand. "Someone you can trust."

QUANTUM LEAP

__ODYSSEY 1-57297-092-8/$5.99
1983: Sam Leaps into troubled but gifted Sean O'Connor. 1999: Al Calavicci has the real Sean O'Connor in "detention" in the Waiting Room—and he's determined to escape.

__TOO CLOSE FOR COMFORT 1-57297-157-6/$5.99
Sam Leaps into a 1990s men's encounter group, only to encounter Al on a mission that could alter the fate of the Quantum Leap project for all time.

__THE WALL 1-57297-216-5/$5.99
Sam Leaps into a child whose destiny is linked to the rise—and fall—of the Berlin Wall.

__PRELUDE 1-57297-134-7/$5.99
The untold story of how Dr. Sam Beckett met Admiral Al Calavicci; and with the help of a machine called Ziggy, the Quantum Leap project was born.

__KNIGHTS OF THE MORNINGSTAR 1-57297-171-1/$5.99
When the blue light fades, Sam finds himself in full armor, facing a man with a broadsword—and another Leaper.

__SEARCH AND RESCUE 1-57297-178-9/$5.99
Sam Leaps into a doctor searching for a downed plane in British Columbia. But Al has also Leapt—into a passenger on the plane.

__PULITZER 1-57297-022-7/$5.99
When Sam Leaps into a psychiatrist in 1975, he must evaluate a POW just back from Vietnam—a Lieutenant John Doe with the face of Al Calavicci.

Based on the Universal Television series created by Donald P. Bellisario

Payable in U.S. funds. No cash accepted. Postage & handling: $1.75 for one book, 75¢ for each additional. Maximum postage $5.50. Prices, postage and handling charges may change without notice. Visa, Amex, MasterCard call 1-800-788-6262, ext. 1, or fax 1-201-933-2316; refer to ad # 530a

Or, check above books   Bill my: ☐ Visa ☐ MasterCard ☐ Amex _____ (expires)
and send this order form to:
The Berkley Publishing Group   Card#_____
P.O. Box 12289, Dept. B   Daytime Phone #_____   ($10 minimum)
Newark, NJ 07101-5289   Signature_____

Please allow 4-6 weeks for delivery.   Or enclosed is my: ☐ check ☐ money order
Foreign and Canadian delivery 8-12 weeks.

Ship to:

Name_____	Book Total $_____
Address_____	Applicable Sales Tax $_____ (NY, NJ, PA, CA, GST Can.)
City_____	Postage & Handling $_____
State/ZIP_____	Total Amount Due $_____

Bill to: Name_____
Address_____   City_____
State/ZIP_____

> "The best dramatic series on television."
> —*New York Daily News*

HOMICIDE

All-new original novels based on characters from NBC's hit TV show.

LIFE ON THE STREET
by Jerome Preisler

Detectives Frank Pembleton and Tim Bayliss must take to the streets to solve a case that threatens to tear the city apart...

___1-57297-227-0/$5.99

VIOLENT DELIGHTS
by Jerome Preisler

Detectives Kellerman and Lewis have a roadside corpse, a guard at Central Maryland Psychiatric Center for the Criminally Insane. And Central Maryland had an escape the previous night...one of the smartest and most vicious hit-men in the history of Baltimore homicide.

___1-57297-340-4/$5.99

Payable in U.S. funds. No cash accepted. Postage & handling: $1.75 for one book, 75¢ for each additional. Maximum postage $5.50. Prices, postage and handling charges may change without notice. Visa, Amex, MasterCard call 1-800-788-6262, ext. 1, or fax 1-201-933-2316; refer to ad #757

Or, check above books	Bill my: ☐ Visa ☐ MasterCard ☐ Amex _____ (expires)
and send this order form to:	
The Berkley Publishing Group	Card#_____
P.O. Box 12289, Dept. B	Daytime Phone #_____ ($10 minimum)
Newark, NJ 07101-5289	Signature_____
Please allow 4-6 weeks for delivery.	Or enclosed is my: ☐ check ☐ money order
Foreign and Canadian delivery 8-12 weeks.	

Ship to:

Name_____	Book Total	$_____
Address_____	Applicable Sales Tax (NY, NJ, PA, CA, GST Can.)	$_____
City_____	Postage & Handling	$_____
State/ZIP_____	Total Amount Due	$_____

Bill to: Name_____

Address_____ City_____
State/ZIP_____

HERCULES

THE LEGENDARY JOURNEYS™

__BY THE SWORD   1-57297-198-3/$5.99

A novel by Timothy Boggs based on the Universal television series created by Christian Williams

Someone has stolen the magical blade and it is up to Hercules to recover it—though he may be in for more than just a fight with ambitious thieves.

__SERPENT'S SHADOW   1-57297-214-9/$5.99

A novel by Timothy Boggs based on the Universal television series created by Christian Williams

Hercules and Iolaus, heed the desperate plea of a small village. A deadly sea monster has been terrorizing the townsfolk, and only the great strength of Hercules can save them.

__THE EYE OF THE RAM   1-57297-224-6/$5.99

A novel by Timothy Boggs based on the Universal television series created by Christian Williams

It is called the Theater of Fun. Run by Hercules's friend Salmoneus, the traveling troupe has dancing girls, jugglers, comedians, and a first-rate magician named Dragar. But Hercules is about to discover that there is a fine line between magic...and sorcery.

__THE FIRST CASUALTY   1-57297-239-4/$5.99

A novel by David L. Seidman based on the Universal television series created by Christian Williams

Someone is posing as Hercules. Someone with superhuman powers of trickery and deception. A certain cloven-hoofed god with a bad attitude...

Copyright © 1998 by MCA Publishing Rights, a Division of MCA, Inc. All rights reserved.

Payable in U.S. funds. No cash accepted. Postage & handling: $1.75 for one book, 75¢ for each additional. Maximum postage $5.50. Prices, postage and handling charges may change without notice. Visa, Amex, MasterCard call 1-800-788-6262, ext. 1, or fax 1-201-933-2316; refer to ad #692

Or, check above books   Bill my: ☐ Visa ☐ MasterCard ☐ Amex _____ (expires)
and send this order form to:
The Berkley Publishing Group Card#_____

P.O. Box 12289, Dept. B   Daytime Phone #_____ ($10 minimum)
Newark, NJ 07101-5289   Signature_____

Please allow 4-6 weeks for delivery.   Or enclosed is my: ☐ check ☐ money order
Foreign and Canadian delivery 8-12 weeks.

Ship to:

Name_____	Book Total $_____
Address_____	Applicable Sales Tax $_____ (NY, NJ, PA, CA, GST Can.)
City_____	Postage & Handling $_____
State/ZIP_____	Total Amount Due $_____

Bill to: Name_____

Address_____City_____
State/ZIP_____

XENA WARRIOR PRINCESS™

__THE EMPTY THRONE   1-57297-200-9/$5.99

A novel by Ru Emerson based on the Universal television series created by John Schulian and Robert Tapert

In a small, remote village, Xena and her protégé, Gabrielle, make a stunning discovery: All of the men in town have disappeared without a trace. They must uncover the truth before it's *their* turn to disappear...

__THE HUNTRESS AND THE SPHINX   1-57297-215-7/$5.99

A novel by Ru Emerson based on the Universal television series created by John Schulian and Robert Tapert

Xena and Gabrielle are asked to rescue a group of kidnapped children, but when they find the kidnapper, Xena realizes that no one is strong enough to defeat it. For who can challenge the power of the almighty Sphinx?

__THE THIEF OF HERMES   1-57297-232-7/$5.99

A novel by Ru Emerson based on the Universal television series created by John Schulian and Robert Tapert

Xena and Gabrielle are framed by Hadrian, who claims to be the son of Hermes, the Sun god. Is Hadrian good or evil? A god's child or a liar?

__PROPHECY OF DARKNESS   1-57297-249-1/$5.99

A novel by Stella Howard based on the Universal television series created by John Schulian and Robert Tapert

Xena and Gabrielle encounter a twelve-year-old seer with a startling prophecy. But more danger awaits. Because according to the prophecy, one of them will not return...

Copyright ©1998 by MCA Publishing Rights, a Division of MCA, Inc. All rights reserved.

Payable in U.S. funds. No cash accepted. Postage & handling: $1.75 for one book, 75¢ for each additional. Maximum postage $5.50. Prices, postage and handling charges may change without notice. Visa, Amex, MasterCard call 1-800-788-6262, ext. 1, or fax 1-201-933-2316; refer to ad #693

Or, check above books and send this order form to:	Bill my: ☐ Visa ☐ MasterCard ☐ Amex _____ (expires)
The Berkley Publishing Group	Card#
P.O. Box 12289, Dept. B	Daytime Phone #_____ ($10 minimum)
Newark, NJ 07101-5289	Signature

Please allow 4-6 weeks for delivery.   Or enclosed is my: ☐ check ☐ money order
Foreign and Canadian delivery 8-12 weeks.

Ship to:

Name	Book Total	$_____
Address	Applicable Sales Tax (NY, NJ, PA, CA, GST Can.)	$_____
City	Postage & Handling	$_____
State/ZIP	Total Amount Due	$_____

Bill to: Name_____

Address_____ City_____
State/ZIP_____